Two week loan

Benthyciad pythefnos

Please return on or before the due date to avoid overdue charges
*A wnewch chi ddychwelyd ar neu cyn y dyddiad a nodir ar eich llyfr os
gwelwch yn dda, er mwyn osgoi taliadau*

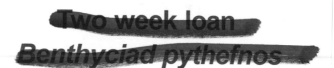

Molecular Biology
of the
Mycobacteria

Edited by

Johnjoe McFadden
Department of Microbiology
University of Surrey
Guildford
UK

SURREY UNIVERSITY PRESS
In association with Academic Press
Harcourt Brace Jovanovich Publishers
London   San Diego   New York
Boston   Sydney   Tokyo   Toronto

Surrey University Press
published by
ACADEMIC PRESS LIMITED
24–28 Oval Road,
London NW1 7DX

United States Edition published by
ACADEMIC PRESS INC.
San Diego, CA 92101

British Library Cataloguing in Publication Data
Molecular biology of the mycobacteria.
1. Mycobacteria
I. McFadden, Johnjoe   II. Series
589.92

ISBN 0–12–483378–0

Typeset by J&L Composition Ltd, Filey, North Yorkshire.
Printed in Great Britain by Galliards (Printers) Ltd, Great Yarmouth, Norfolk.

Contents

Contributors

C. Abou-Zeid Department of Microbiology, University College and Middlesex School of Medicine, Riding House St, London W1P 7PN, UK

B. R. Bloom Department of Microbiology and Immunology, Albert Einstein College of Medicine, Bronx, NY 10461, USA

J. E. Clark-Curtiss Departments of Molecular Microbiology and Biology, Washington University, St Louis, MO 63130, USA

M. J. Colston Laboratory for Leprosy and Mycobacterial Research, National Institute for Medical Research, The Ridgeway, Mill Hill, London NW7 1AA, UK

J. T. Crawford Medical Research Service, John L. McClellan Memorial Veterans Hospital, and the Department of Microbiology and Immunology, University of Arkansas for Medical Sciences, 4300 W 7th, Little Rock, AR 72205, USA

J. W. Dale Department of Microbiology, University of Surrey, Guildford, Surrey GU2 5XH, UK

J. O. Falkinham, III Department of Biology, Virginia Polytechnic Institute and State University, Blacksburg, Virginia 24061, USA

T. Garbe MRC Tuberculosis and Related Infections Unit, Royal Postgraduate Medical School, Hammersmith Hospital, London W12 0HS, UK

B. Gicquel Unité de Génie Microbiologique, Département des Biotechnologies, URA 209 du CNRS, Institut Pasteur, 28 rue du Docteur Roux, 75015 Paris, France

W. R. Jacobs, Jr Department of Microbiology and Immunology, Albert Einstein College of Medicine, Bronx, NY 10461, USA

Z. Kunze Department of Microbiology, University of Surrey, Guildford, Surrey GU2 5XH, UK

R. Lathigra MRC Tuberculosis and Related Infections Unit, Royal Postgraduate Medical School, Hammersmith Hospital, London W12 0HS, UK

C. Martin Unité de Génie Microbiologique, Département des Biotechnologies, URA 209 du CNRS, Institut Pasteur, 28 rue du Docteur Roux, 75015 Paris, France

J. McFadden Department of Microbiology, University of Surrey, Guildford, Surrey GU2 5XH, UK

A. Patki Department of Microbiology, University of Surrey, Guildford, Surrey GU2 5XH, UK

M. Ranes Unité de Génie Microbiologique, Département des Biotechnologies, URA 209 du CNRS, Institut Pasteur, 28 rue du Docteur Roux, 75015 Paris, France

P. Seechurn Department of Microbiology, University of Surrey, Guildford, Surrey GU2 5XH, UK

S. B. Snapper Department of Microbiology and Immunology, Albert Einstein College of Medicine, Bronx, NY 10461, USA

J. E. R. Thole Immunohaematology and Blood Bank, University Hospital Leiden, PO Box 9600, 2300 RC Leiden, The Netherlands

D. Young MRC Tuberculosis and Related Infections Unit, Royal Postgraduate Medical School, Hammersmith Hospital, London W12 0HS, UK

R. A. Young Whitehead Institute for Biomedical Research, Nine Cambridge Center, Cambridge, MA 02142, USA, and Department of Biology, Massachusetts Institute of Technology, Cambridge, MA 02139, USA

R. van der Zee National Institute of Public Health and Environmental Protection, Laboratory of Bacteriology, PO Box 1, 3720 BA Bilthoven, The Netherlands

Preface

This volume is the first comprehensive account of the enormous progress that has been made in the five years since first reports of cloning of mycobacterial genes were published. We trust that this volume will be invaluable to the increasing numbers of scientists now working in this field and to myco-bacteriologists generally, including clinicians concerned with epidemiology and new diagnostics for mycobacterial diseases. We also hope that those with an interest in bacterial genetics, microbial pathogenicity or DNA probes will find much of value in the following chapters.

Mycobacterial genetics is now one of the most rapidly advancing fields in bacterial genetics. However, mycobacteria are more than an interesting genetic system. Mycobacterial disease remains one of the most intractable world health problems. This may seem surprising to many of us who live in the West, where thanks largely to chemotherapy and vaccination the 'white plague' of tuberculosis has been virtually eradicated and leprosy is considered to be a disease of the distant past. Tragically however, tuberculosis kills 2–3 million people each year and leprosy continues to afflict 20 million world-wide. In addition, the spectre of AIDS with concomitant mycobacterial infections, is bringing tuberculosis back into western cities and seriously threatens the already overburdened health services of the developing world. Reasons why these diseases remain a problem nearly 50 years since the introduction of BCG vaccine and effective chemotherapy for tuberculosis, are of course manifold; but certainly include inconsistency of vaccine efficacy and lengthy treatment regimes. Greater understanding of mechanisms of mycobacterial pathogenesis and immunity will certainly be required in order that more effective vaccination and treatment programmes be developed. Much of mycobacterial molecular biology is now concerned with addressing these issues.

Fortunately mycobacteria have been amongst the most intensely studied bacterial pathogens. A wealth of information has been painstakingly gathered on the immunopathology, and physiology of mycobacteria. Initially myco-bacterial genetics lagged behind other fields, chiefly because of the frustra-tions of working with organisms that seemed resistant to techniques used for genetic manipulation of other bacteria; and in addition, took several weeks to grow, or like the leprosy bacillus, could not be grown except in experimental animals. This volume is, however, proof that this is no longer the case; and that mycobacterial genetics has already contributed significantly to our understanding of microbial pathogenesis and immunity. In addition, DNA probes are providing new tools for rapid and accurate mycobacterial disease diagnosis. It is hoped that the work described herein will form the foundation for even more exciting advances in the years to come; and that a detailed knowledge of molecular biology will be integrated with information from other fields, to achieve a comprehensive understanding of the mycobacteria.

J. McFADDEN

1 Protein antigens: structure, function and regulation

Douglas Young, Thomas Garbe, Raju Lathigra and Christiane Abou-Zeid*

MRC Tuberculosis and Related Infections Unit, Royal Postgraduate Medical School, Hammersmith Hospital, London W12 0HS, UK
* Department of Microbiology, University College and Middlesex School of Medicine, Riding House Street, London W1P 7PN, UK

1 INTRODUCTION

Identification and characterization of individual components of mycobacteria which are involved in interactions with the immune system has provided a major focus for research in mycobacterial diseases. Such studies were initiated in order to assist in understanding of the molecular mechanisms involved in the immune response to mycobacteria, and with the potential practical benefits of identifying candidate molecules for use as 'subunit vaccines' and as reagents for immunodiagnostic tests. In addition, pursuit of these immunological goals has stimulated attempts to apply molecular biology techniques to mycobacteria, and genes coding for many of the prominent protein antigens have been cloned and sequenced. Detailed characterization of these antigens has shown that they include several proteins with interesting functional and regulatory properties and it is likely that further study of such antigens will contribute to improved understanding of fundamental aspects of mycobacterial physiology and metabolism. In this chapter, the protein antigens will be reviewed firstly in the context of the immunological studies during which they were isolated, and then from the perspective of their biochemical function within the mycobacteria.

1.1 The immune response to mycobacteria

The encounter between a microbial pathogen and a potential animal host involves a complex series of interactions with the outcome — resistance, infection, or disease — being dependent both on the host response and on intrinsic properties of the pathogen. The importance of host factors is particularly striking in the case of mycobacterial diseases and, in contrast to many other bacterial diseases, the process of mycobacterial infection has been studied primarily from the perspective of the host rather than of the pathogen.

The key role of the immune response to mycobacteria is clearly illustrated during infection with *Mycobacterium leprae*. Most people living in leprosy endemic areas are exposed to *M. leprae*, but a healthy immune response is sufficient to protect against disease for the vast majority of individuals. Amongst those individuals who do develop disease a wide spectrum of clinical symptoms is seen, reflecting quite distinct differences in their immune response to the pathogen (Ridley and Jopling, 1966) (Figure 1.1). At one end of the spectrum, patients with lepromatous leprosy fail to mount a cell-mediated immune response to *M. leprae* and unrestricted bacterial multiplication occurs, with lesions containing up to 10^9 acid-fast bacilli per gram of tissue. A contrasting situation is seen in tuberculoid leprosy, with a strong cellular immune response restricting bacterial growth (few, if any, bacteria are seen in the lesions) but at the expense of causing severe immunopathological damage to surrounding nerve tissues.

It can be proposed, therefore, that there are two sides to the immune response to mycobacteria. On the one hand it has a protective anti-mycobacterial function, but at the same time it also has a pathological function in the disease process. Understanding of the basic mechanisms of the immune response might allow development of clinical interventions leading to preferential stimulation of protective function, or inhibition of pathological function. From the point of view of antigen characterization, it might be anticipated that this could be accomplished by isolating individual antigens capable of inducing different forms of immune activation. In order to assess this possibility it is necessary to review current ideas concerning mechanisms involved in the immune response to mycobacteria.

1.2 Protection, pathogenesis and T-cell subsets

During infection mycobacteria are generally found inside host phagocytic cells. Their intracellular location, and a lipid-rich cell wall and protective capsule, inhibit attack of mycobacteria by antibody and complement. Although antibodies are induced during infection, therefore, it is the cell-mediated immune system which plays the dominant role in mycobacterial

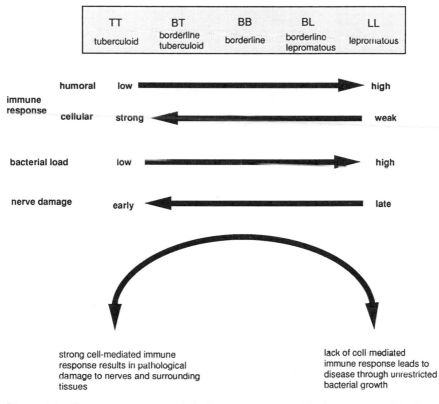

Figure 1.1 Immune response and the leprosy spectrum. The importance of the host immune response in determining the outcome of mycobacterial infection is clearly illustrated by the leprosy spectrum. A wide variety of clinical manifestations is seen, reflecting individual differences in the cell-mediated immune response to *M. leprae*.

immunity (Hahn and Kaufmann, 1981). In the classic concept of cell-mediated immunity to intracellular pathogens developed by Mackaness and co-workers in the 1960s, it is envisaged that regulatory T lymphocytes are triggered by specific antigen recognition and that these cells can go on to activate macrophages in order to enhance their non-specific microbicidal function (Mackaness, 1969). *In vitro* experiments have indicated that macrophage killing of pathogenic mycobacteria is a relatively inefficient process, however, and current concepts of protective immunity include an extension of the straightforward Mackaness model (Kaufmann, 1988). A role has been proposed for cytotoxic T cells in release of mycobacteria from the 'protected' environment of a non-microbicidal cell, or of a phagocyte which has been rendered refractory to activation. The free mycobacteria are then available

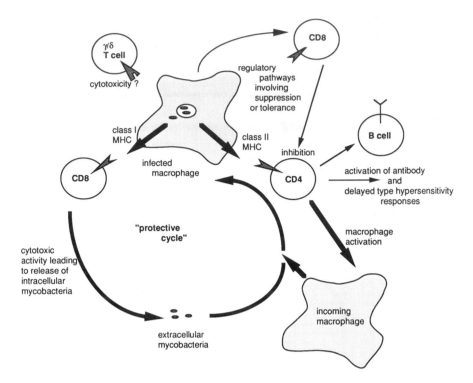

Figure 1.2 Multiple functions involved in the immune response to mycobacteria. Infection with a mycobacterial pathogen results in stimulation of multiple immune functions. Activation of macrophages by antigen-specific CD4 helper T cells is thought to play a major role in mycobacterial killing, but it may be necessary to co-ordinate such activity with lysis of certain infected cell populations by cytotoxic T cells. T cells with a γ/δ receptor may be particularly important during the early stages of the immune response, and a role for antigen-specific suppressor T cells has been proposed in lepromatous leprosy.

An appropriate balance between the different immune functions may be important for protective immunity, while excessive activation of one or more pathways may result in immunopathology.

for phagocytosis by newly activated incoming macrophages, which may in turn be lysed by cytotoxic cells if they fail to control the pathogen (Figure 1.2).

The model outlined in Figure 1.2 indicates that protective immunity arises from the closely coordinated activation of multiple immune functions. Uncoordinated activation of the same functions — resulting in excessive macrophage activation, or excessive cytotoxicity — could account for the pathological consequences of the immune response during disease. The concept of protective immunity as a complex balance of immune functions is important with regard to the potential evaluation of individual antigens. While it may be possible to use a single antigen to tilt the balance of the immune response in one way or another, it seems probable that the overall balance will be the result of multiple interactions with multiple antigens.

Antigens which take part in T-cell recognition must first be unfolded and partially degraded ('processed') within an antigen presenting cell (APC). Antigen fragments bind to a protein coded by genes from the major histocompatibility complex (MHC) and the antigen-bound MHC molecule is then displayed on the surface of the APC in a form suitable for recognition by a specific receptor on T cells (reviewed in Schwartz, 1985). Two distinct processing pathways are thought to exist, with antigens present in the cytoplasm of the APC being presented with class I MHC molecules, and antigens from phagocytic vesicles being linked to class II MHC (Germain, 1986). An important functional consequence of the different processing pathways is that binding to the antigen-MHC class I complex is enhanced by the presence of a CD8 protein on the T-cell surface, while a CD4 protein promotes binding to MHC class II antigen complexes. CD8[+] T cells generally express a cytotoxic function, while CD4[+] cells are predominantly 'helper' cells which function in macrophage and B-cell activation. While there may be some exceptions to this division (a cytotoxic activity of some CD4[+] cells has been reported, e.g. Ottenhoff et al., 1988), the implication is that the location of an antigen within the APC will influence its role in the immune response. In addition to binding of the antigen receptor, additional signals — generally provided by the APC in the form of a cell–cell interaction, or delivery of a soluble cytokine — are required for T-cell activation (Mueller et al., 1989). It is possible that the precise function of the activated T cell is influenced by the nature of these secondary signals and there may not, therefore, be a simple relationship between recognition of a particular antigen and induction of a particular form of immune activation.

A further functionally distinct lymphocyte subset which has frequently been discussed in the context of mycobacterial immunology is the CD8[+] suppressor cell. The concept of a lymphocyte subset with the ability to suppress immune responses was introduced in order to account for in vivo tolerance induction experiments, and was considered to play an important role in regulation of the immune system and in preventing responses to 'self' antigens. Activation of suppressive pathways provides an attractive explanation for the unresponsiveness associated with lepromatous leprosy and a considerable effort has gone into assessing the significance of suppressor cells in leprosy (Bloom and Mehra, 1984). However, a variety of alternative explanations have been put forward to account for some of the phenomena associated with suppression (cytotoxicity, scavenging of growth factors) and a lack of progress in cloning of the relevant cell types has raised questions as to the general applicability of the suppressor-cell concept (see Schwartz, 1989 for review).

In identifying genes for the antigen-specific receptor of T lymphocytes it was found that, in addition to the commonly used alpha and beta chains, two related genes (gamma and delta) coded for a similar receptor molecule (Raulet, 1989). The gamma–delta receptor is present on developing lymphocytes and on a small percentage of mature T cells — particularly at epithelial sites. The function of these gamma–delta T cells is poorly understood at present, though they show antigen-specific cytotoxic activity *in vitro* and a role in elimination of abnormal or infected cells has been proposed (Janeway *et al.*, 1988). It has recently been shown that the gamma–delta T-cell population is enriched in leprosy lesions and in the primary immune response to *M. tuberculosis* (Janis *et al.*, 1989; Modlin *et al.*, 1989), and the role of these cells in anti-mycobacterial immunity is an active area of research interest.

2   IDENTIFICATION AND CLONING OF ANTIGENS

2.1   Which antigen?

Considering the complexity of the immune response described above, it seems unlikely that any single antigen will hold the key to understanding mycobacterial immunity. It is useful therefore to adopt several different criteria in selecting individual antigens for detailed immunological analysis.

2.1.1   Immunodominance

It is likely that immunodominant antigens — those which are the most frequent targets of immune recognition — will play a correspondingly important role in determining the overall balance of the immune response. Immunodominance will be a reflection of a combination of host factors — the immune system may have a bias for particular antigens — and inherent features of the protein, including its structural properties and its relative concentration. In addition, different analytical techniques may also impose differences in apparent immunodominance. Western blotting, which involves antibody interactions with denatured proteins may, for example, reveal a different pattern to that seen by immunoprecipitation assays which depend on binding to native structures. A variety of attempts have been made to identify patterns of immunodominance in antibody and cell-mediated responses to mycobacteria.

Analysis of the antibody response of inbred mouse strains infected with

M. tuberculosis or immunized with mycobacterial extracts indicates that there is a hierarchy by which particular proteins are recognized in preference to others (Ivanyi and Sharp, 1986; Ljungqvist *et al.*, 1988; Huygen *et al.*, 1988a; Brett and Ivanyi, submitted for publication). The pattern of immuno-dominance differs between different mouse strains with some of the differences being associated with particular MHC haplotypes. Similarly, in humans with mycobacterial infections, Western blot analysis often reveals a striking immunodominance in the pattern of antibody recognition (Klatser *et al.*, 1984; Rumschlag *et al.*, 1988; Vega-Lopez *et al.*, 1988; Espitia *et al.*, 1989). Comparison of Western blot profiles from different patients shows that, while a few common bands can be picked out, quite distinct individual variations exist in the apparent immunodominance of different proteins. The multiple factors which underly this variation probably include genetic differences (MHC and non-MHC), previous history of infection, and differences in clinical progression of the disease.

Experiments involving fusion of spleens from Balb/c mice immunized with cell extracts and culture filtrates of *M. tuberculosis* or *M. leprae* resulted in generation of panels of monoclonal antibodies recognizing a limited set of proteins — presumably reflecting the pattern of immunodominance in this mouse strain (Engers *et al.*, 1985, 1986) (Table 1.1). These 'standard' monoclonal antibodies have been extensively used as probes for antigen identification in diagnostic assays and in screening recombinant DNA libraries. Antibody responses to the proteins identified by monoclonal antibodies are detectable in the majority of patients with multibacillary infections, suggesting some form of immunodominance for these antigens at a population level (Ivanyi *et al.*, 1988). However, it should be emphasized that these antigens represent only a subset of the proteins which take part in human immune responses to mycobacteria and, indeed, further experiments have shown that many additional proteins are also recognized by monoclonal antibodies generated in Balb/c mice (e.g. Damiani *et al.*, 1988; Ljungqvist *et al.*, 1988).

Immunodominance in the T-cell response has been investigated using lymphocyte proliferation assays. Proliferation experiments with fractionated bacterial extracts immobilized in the form of nitrocellulose particles again show striking differences between different individuals with no clear patterns associated with different forms of disease (Converse *et al.*, 1988; Mendez-Samperio *et al.*, 1989; Filley *et al.*, 1989). As in the case of the antibody response, these differences probably reflect a complex array of host factors including genetic, environmental and clinical influences. Individual differences in relative immunodominance are also reflected in an elevated frequency of responses to particular antigens as judged by T-cell cloning experiments (Emmrich *et al.*, 1985; Mustafa *et al.*, 1986; Lamb *et al.*, 1986;

Table 1.1 Monoclonal Antibodies Characterized in WHO Workshops.

Antibody code	Antigen mol.wt (kD)	Species used for immunization	Cross-reaction with other mycobacteria
51A	71	M. tuberculosis	yes
IIIE9, IVD8		M. leprae	no
IIH9, ML30	65	M. leprae	yes
IIC8, Y1.2			
TB78		M. tuberculosis	limited
TB71, TB72 HYT28	38	M. tuberculosis	no
F47.9	36	M. leprae	no
ML04	35	M. leprae	no
D2D	23*	M. leprae	yes
TB23, HYT6 F29.47	19	M. tuberculosis	yes
L5	18	M. leprae	no
F24.2, TB68 F23.49	14	M. tuberculosis	limited
ML06	12	M. leprae	no
SA12	12	M. bovis BCG	no

* The molecular weight of the antigen was estimated as 23 kD in *M. tuberculosis* and as 28 kD in *M. leprae*.
Details of the WHO Workshops, and references to the generation and characterization of the monoclonal antibodies, are provided in Engers *et al.*, 1985 and 1986.

Ottenhoff *et al.*, 1986a; Oftung *et al.*, 1987). It has recently been demonstrated that proteins associated with the cell wall of mycobacteria provide a particularly potent stimulus for T-cell proliferation (Melancon-Kaplan *et al.*, 1988; Hunter *et al.*, 1989; Mehra *et al.*, 1989; Barnes *et al.*, 1989). Further analysis with isolated proteins will show whether this immunodominance is an intrinsic property of the proteins themselves, or if it is their physical association with the mycobacterial cell wall which leads to enhanced immunogenicity.

2.1.2 Species specificity

A second criterion which has been used in selection of antigens is their relative degree of species-specificity. Distinction between the immune response to an infecting mycobacterial pathogen and cross-reactive background responses to environmental bacteria is of importance in development of diagnostic tests and the criterion of species-specificity has been particularly emphasized in studies with diagnostic goals. Monoclonal antibody screens have therefore often given preference to identification of proteins which have species-specific antibody epitopes. A species-specific epitope can be generated simply by the alteration of one or two amino acid residues in a protein, however, and antigens identified by specific antibodies have often been shown to carry additional cross-reactive antibody or T-cell determinants (e.g. Young *et al.*, 1986). T-cell clones have also been used to detect species-specific T-cell epitopes on mycobacterial proteins (e.g. Lamb and Young, 1987).

In contrast to the original emphasis on species-specific epitopes, there has been a growing interest in the possible significance of highly conserved determinants on mycobacterial antigens — particularly in the case of antigens which belong to heat shock protein families (see below). It has been suggested that foreign antigens which share extensive sequence identity with 'self' proteins may be important in triggering regulatory pathways involving tolerance or suppression (Young *et al.*, 1988b). Conversely, breakdown in such regulation could lead to autoimmune reactions which may contribute to immunopathological damage (see Thole, this volume).

The species-specificity of some of the anti-mycobacterial monoclonal antibodies is shown in Table 1.1.

2.1.3 Secreted proteins

Proteins which are secreted from mycobacteria during growth may become available for immune recognition at an early stage of the infection — prior to the release of intracellular components from killed organisms. Triggering of an early response may make an important contribution to protective immunity and recognition of secreted proteins may therefore be preferentially associated with protection (Rook *et al.*, 1986; Orme, 1987). As discussed above, subcellular location may be an important factor in determining the pathway by which an antigen is processed and presented and it is possible that secreted antigens differ from internal components in their ability to penetrate different intracellular compartments. A number of studies have focused on identification and characterization of proteins released from mycobacteria during *in vitro* culture.

Autolysis and cell-wall damage can contribute to release of proteins into the medium during growth of mycobacteria and several criteria have been used in attempts to distinguish proteins which are being actively secreted by intact live bacilli. These have been based on quantitative comparison of culture medium and cell extracts (Harboe *et al.*, 1986; Wiker *et al.*, 1986), and on the detection of radiolabelled proteins in short-term cultures (Abou-Zeid *et al.*, 1988a). While these approaches represent a useful primary screen, sequence analysis of cloned genes now provides an important means of clearly identifying proteins which are designed for export from the cell — antigens which accumulate in the culture medium can be separated into two classes, distinguished by the presence or absence of a signal peptide (see Section 4.2). It is proposed that the term 'secreted proteins' should be reserved for those proteins for which a signal peptide mediates their active transport across the cell membrane.

A major contribution towards characterization of antigens in culture filtrates has been provided by the technique of crossed immunoelectrophoresis (CIE) by which a large number of mycobacterial proteins have been identified on the basis of their interaction with hyperimmune antisera (Closs et al., 1980, Wiker et al., 1988). Several of the proteins identified by CIE have been further characterized by N-terminal amino acid sequencing and preparation of monospecific antisera. Some of the major secreted antigens studied in this way are listed in Table 1.2, and a comparison of proteins in early and late culture supernatants by SDS polyacrylamide gel electrophoresis is illustrated in Figure 1.3.

Table 1.2   Major Culture Filtrate Antigens Defined by CIE.

CIE reference code	Alternative name	Mol.wt (kD)	Species-specificity
(a) Proteins with a signal peptide			
Antigen 85 85A	P32		
complex 85B	MPB59, α antigen	28–32	broad
85C			
Antigen 64	MPB64	23	specific to M. tuberculosis and M. bovis BCG
Antigen 70 (identical)	MPB 70	18	specific
Antigen 80 (N-terminus)	MPB 80		to M. bovis
Antigen 78	US–Japan Antigen 5	38	specific to M. tuberculosis
	PhoS		and M. bovis BCG
(b) Proteins without a signal peptide			
Antigen 63	DnaK	71	broad
Antigen 82	GroEL	65	broad

Data from: Closs et al., 1980; Harboe et al., 1986; Wiker et al., 1986, 1988; Abou-Zeid et al., 1988a; Huygen et al., 1988a, b; Matsuo et al., 1988; Radford et al., 1988; Borremans et al., 1989; Terasaka et al., 1989; and Yamaguchi et al., 1989.

2.2   Gene cloning

Having identified antigens on the basis of the criteria listed above, the application of molecular biology techniques has played an important role in

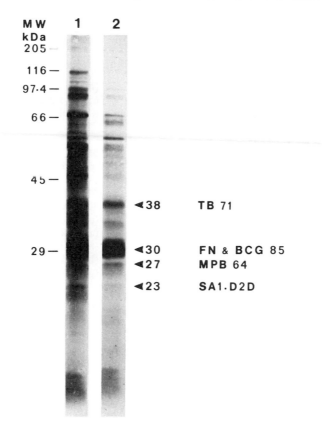

Figure 1.3 Analysis of secreted proteins by SDS-PAGE. Proteins in supernatants from 21 day (lane 1) and 3 day (lane 2) cultures of *M. tuberculosis* were separated by SDS-PAGE on 12.5% acrylamide gels, transferred to nitrocellulose membranes by electroblotting, and stained with Aurodye, a colloidal gold solution. The reactivity of monoclonal antibodies TB71 and D2D, of anti-MPB64 and anti-BCG85 antisera, and of fibronectin on immunoblots of culture supernatants of *M. tuberculosis* are shown on the right, and the molecular weights of the protein antigens detected are indicated.

Although many proteins are found in the supernatant of 'old' cultures, only a few proteins are accumulated during the early stages of growth.

allowing a detailed investigation of the proteins. The most widely applied approach to cloning of the genes for mycobacterial antigens has been the use of the lambda gt11 system (Young *et al.*, 1985a, b). This is based on the expression of antigens as beta-galactosidase fusion proteins which are detected by immunoblotting of phage plaques on an *E. coli* lawn, and is ideally suited for proteins which are defined by antibody probes. The

Table 1.3 Cloned Genes for Mycobacterial Antigens.

Mol. wt (kD)	M. tuberculosis	M. bovis	M. leprae	Refs
70	71 kD antigen (P) (Mab 51A)		70 kD antigen (P) (Mab L7)	1–4
65	65 kD antigen (C) (multiple Mabs)	64 kD antigen (C) (polyclonal antiserum)	65 kD antigen (C) (multiple Mabs)	1, 2, 5–8
38	38 kD antigen (C) (Mab HYT28)			9
36			36 kD antigen (C) (multiple Mabs)	10
35	35 kD antigen (C) (Mab H63.1)			11, 12
30–31	P32 (C) (polyclonal serum)	α antigen (C) (oligonucleotide)	30/31 kD antigen (polyclonal serum)	13–15
28			(a) 28 kD antigen (C) (Mabs B11H, D2D)	2, 16
			(b) 28 kD antigen (C) (patient's serum)	17
23	23 kD antigen (P) (Mab D2D)			18
23		MPB 64 (C) (oligonucleotide)		19
19	19 kD antigen (C) (Mab TB23)			1, 20
18		MPB70 (C) (polyclonal antiserum, oligonucleotide)		21, 22
18			18 kD antigen (C) (Mab L5)	2, 23, 24
14	14 kD antigen (Mab TB68)			1, 25
12	BCG-a (C) (Mab SA12)	BCG-a, MPB57 (C) (oligonucleotide)		26–28
12			12 kD antigen (Mab ML10)	18

Mycobacterial antigens for which genes have been cloned are listed by their molecular weight, along with an indication of the method used for identification of the gene (antibody screen or oligonucleotide hybridization). The designation in parenthesis shows the availability of partial (P) or complete (C) sequence information for the gene. Key to references: (1) Young et al., 1985a; (2) Young et al., 1985b; (3) Young et al., 1988a; (4) Garsia et al., 1989; (5) Shinnick, 1987; (6) Thole et al., 1985; (7) Thole et al., 1987; (8) Mehra et al., 1986; (9) Andersen and Hansen, 1989; (10) R. Hartskeerl and J. Thole, personal communication; (11) Cohen et al., 1987; (12) D. Vismara and G. Damiani, personal communication; (13) Borremans et al., 1989; (14) Matsuo et al., 1988; (15) C. Abou-Zeid, unpublished data; (16) H. Thangaraj and M. J. Colston, personal communication; (17) Cherayil and Young, 1988; (18) Y. Zhang, R. Lathigra and D. Young, unpublished data; (19) Yamaguchi et al., 1989; (20) Ashbridge et al., 1989; (21) Radford et al., 1988; (22) Terasaka et al., 1989; (23) Booth et al., 1988; (24) Nerland et al., 1988; (25) Young et al., 1987; (26) Baird et al., 1988; (27) Shinnick et al., 1989; (28) Yamaguchi et al., 1988.

monoclonal antibodies described above have been particularly effective probes in that they react with single gene products, often binding to epitopes which are conserved in the denatured polypeptide. The genes for most of the antigens identified by monoclonal antibodies to *M. tuberculosis* and *M. leprae* have been cloned using this approach (Table 1.3). In addition to the expected fusion protein products, several antigens have been found as 'free' proteins in lambda gt11 lysogens, indicating that some mycobacterial transcription and translation signals are functional in *E. coli*. Similarly, some mycobacterial antigens have been detected in expression systems which do not rely on the presence of a strong exogenous promoter (Thole *et al.*, 1985). Polyclonal antisera — from hyperimmune animals, or from patients — have also been used to screen lambda gt11 libraries in many laboratories and, in a few cases, the antigens recognized have been characterized (Young *et al.*, 1987; Radford *et al.*, 1988; Cherayil and Young, 1988). The possibility of screening lambda gt11 libraries with T-cell clones has also been investigated (Lamb *et al.*, 1988; Mustafa *et al.*, 1988).

An alternative approach has been to determine the N-terminal sequence of purified protein antigens and to screen recombinant DNA libraries with appropriately designed oligonucleotide probes (Matsuo *et al.*, 1988). The characteristic codon usage pattern of mycobacteria (high GC percentage in the third base position) has assisted in this approach, which has been successfully employed for several antigens of *M. bovis* BCG (Table 1.3). The extensive work carried out on biochemical fractionation of secreted antigens (e.g. Harboe *et al.*, 1986) has been particularly important in permitting application of the oligonucleotide approach to cloning the genes for these proteins.

Table 1.3 lists mycobacterial antigens for which the genes, or parts of the genes, have been cloned. In many cases the nucleotide sequence of the gene has been determined.

3   IMMUNOLOGICAL ANALYSIS OF DEFINED ANTIGENS

3.1   T-cell recognition

A major objective of antigen characterization has been to determine if the defined proteins play a role in the cell-mediated immune response to mycobacteria, and whether or not recognition of different antigens induces different forms of immune activation.

Cellular immune responses to defined antigens have been tested using either proteins purified from mycobacterial cultures or expressed as recombinant products in *E. coli* in assays based on induction of delayed type hypersensitivity or T-cell proliferation (Minden *et al.*, 1984; Emmrich *et al.*, 1985; Mustafa *et al.*, 1986; Ottenhoff *et al.*, 1986a; Lamb *et al.*, 1986; Young *et al.*, 1986; Britton *et al.*, 1986; Oftung *et al.*, 1987; de Bruyn *et al.*, 1987; Huygen *et al.*, 1988b). In every case it has been possible to demonstrate T-cell recognition, and a general conclusion which can be drawn from such studies is that a large number of mycobacterial proteins can take part in immune recognition during infection or following experimental immunization. Quantitative analysis of lymphocyte responses has been carried out only for the murine response to the 65 kD antigen. In this case, limiting dilution experiments showed that up to 20% of the mycobacteria-reactive T cells induced in response to immunization with *M. tuberculosis* are directed to this single protein — suggesting a truly immunodominant role for the 65 kD antigen (Kaufmann *et al.*, 1987). Comparative studies with other antigens may allow establishment of a hierarchy of relative immunodominance for each protein — though such a hierarchy may well be quite distinct for individuals with different genetic and clinical backgrounds.

Several approaches can be proposed in order to determine whether different antigens stimulate different forms of immune activation. Vaccination with defined antigens in animal models of disease provides a direct approach to analysis of protective immunity and this is an active area of current research (see Colston, this volume). Alternative approaches which allow analysis of the role of particular antigens in human immune responses involve (a) comparison of the repertoire in infected and immune individuals, and (b) detailed examination of differential stimulation of lymphocyte subsets *in vitro*.

(a) Six antigens from *M. tuberculosis* (71 kD, 65 kD, 38 kD, 30/31 kD, 19 kD and 12 kD) and the 18 kD antigen from *M. leprae* have been tested for their ability to induce proliferative responses in patients and in healthy controls (Young *et al.*, 1986; Britton *et al.*, 1986; Munk *et al.*, 1988; Huygen *et al.*, 1988b; Dockrell *et al.*, 1989). In each case, a response was found in both groups indicating that recognition of these antigens is not exclusively associated either with protective or with pathogenic forms of immune activation. Considerable individual variations in T-cell proliferation were found, however, and a generally higher response to the 18 kD antigen of *M. leprae* in household contacts as compared to tuberculoid patients led to a suggestion that some degree of preferential association with protection may occur (Dockrell *et al.*, 1989). It is likely that this type of repertoire analysis will be carried out more frequently as additional recombinant antigens become available in large amounts from overexpression vectors. A useful approach to analysing the results may be to compare the response of individuals to different groups of antigens. It might be interesting, for example, to determine the relative response to secreted versus cytoplasmic antigens for different patient groups.

(b) *In vitro* stimulation of lymphocyte proliferation generally involves antigen recognition by CD4$^+$ T cells which are presumed to function primarily in B cell and macrophage activation pathways. Recognition by functionally distinct T-cell subsets has been demonstrated for some antigens. The 65 kD antigen, for example, stimulates cytotoxic responses involving both CD4$^+$ and CD8$^+$ T cells (Ottenhoff *et al.*, 1988; Munk *et al.*, 1989; Koga *et al.*, 1989), and recognition of the same antigen by T cells carrying the gamma–delta receptor has also been reported (Holoschitz *et al.*, 1989; O'Brien *et al.*, 1989; Haregewoin *et al.*, 1989). The 71 kD antigen of *M. tuberculosis* is also recognized by both CD4$^+$ and CD8$^+$ cells (Rees *et al.*, 1988). The 36 kD antigen of *M. leprae* activates CD4$^+$ helper cells, and also a subset of CD8$^+$ T lymphocytes which are able to suppress proliferation of the CD4$^+$ cells (Ottenhoff *et al.*, 1986b). The studies which have been carried out so far regarding antigen recognition by different lymphocyte subsets are insufficient to allow any general conclusions about links between particular proteins and different forms of immune activation. They do, however, illustrate the fact that a single polypeptide does have the capacity to stimulate multiple immune functions, thereby emphasizing the importance of defining single antigenic determinants involved in immune recognition.

3.2 Epitope mapping

The functional 'unit' involved in T-cell recognition appears to be a peptide fragment located in the antigen binding site of an MHC molecule. It can be argued, therefore, that a true dissection of the immune response by means of antigen fractionation must involve attempts to define such antigen–MHC complexes. Mapping of T-cell epitopes by a combination of recombinant DNA and synthetic peptide technologies has been enthusiastically pursued using the 65 kD antigen as a model system and is reviewed in detail by Thole in this volume. The definition of peptide epitopes recognized in the context of different MHC haplotypes allows a further level of detail in the analysis of repertoire and may ultimately identify determinants which are uniquely associated with particular forms of immune activation.

3.3 Summary

An extensive body of research has resulted in the identification and cloning of genes for a variety of mycobacterial antigens differing in their subcellular location and in their relative content of species-specific and cross-reactive epitopes. There are clearly quantitative differences in the immune response which the defined antigens induce in different individuals but data available at present are insufficient to determine whether or not different proteins are associated with qualitative differences in immune activation.

4   BIOCHEMICAL AND GENETIC ANALYSIS OF DEFINED ANTIGENS

In the remaining portion of this chapter, the proteins originally identified on the basis of their interaction with the host immune system will be reviewed with respect to their function from the perspective of the mycobacterial parasite.

4.1   Heat shock proteins

On the basis of sequence conservation, antigenic cross-reactivity, and regulation of expression, several of the major antigens of mycobacteria have been identified as members of heat shock protein families (Table 1.4) (Young *et al.*, 1988a; Shinnick *et al.*, 1988; Baird *et al.*, 1988; Nerland *et al.*, 1988; Shinnick *et al.*, 1989; Mehlert and Young, 1989; Garsia *et al.*, 1989).

Table 1.4   Mycobacterial Antigens which Belong to Heat-shock Protein Families.

hsp family	E. coli	M. tuberculosis	M. leprae	*Refs*
hsp 70	DnaK	71 kD antigen	70 kD antigen	Young *et al.*, 1988a
				Garsia *et al.*, 1989
				Mehlert and Young, 1989
hsp 60	GroEL	65 kD antigen	65 kD antigen	Young *et al.*, 1988a
				Shinnick *et al.*, 1988
—	GroES	12 kD antigen		Baird *et al.*, 1988
		(BCG-a, MPB57)		Shinnick *et al.*, 1989
low mol. wt	—		18 kD antigen	Nerland *et al.*, 1988

Exposure of cultured cells or organisms to an increase in temperature results in induction of the synthesis of a set of 'heat-shock' proteins (reviewed by Lindquist, 1986; Lindquist and Craig, 1988). The heat-shock response, and the proteins themselves, are highly conserved throughout nature with more than 50% sequence identity between corresponding proteins from *E. coli* and man. Although the heat-shock proteins were originally identified by their specific role in response to temperature, it is now known that they can be induced by a wide variety of environmental stress stimuli — including exposure to alcohol, oxidative radicals, and toxic metal ions — and the more general term 'stress proteins' is often used interchangeably with 'heat-shock proteins'. Members of stress protein families are also present at high concentrations in 'unstressed' cells and are essential for normal cell growth under all conditions.

In eukaryotic cells the predominant heat-shock proteins belong to a group of proteins with molecular weights around 70 kD — referred to as the hsp70 gene family. Bacteria express a single hsp70 protein — the product of the *dnaK* gene in *E. coli* — and this corresponds to the 71 kD and 70 kD proteins identified as major antigens in *M. tuberculosis* and *M. leprae* respectively (Young *et al.*, 1988a; Garsia *et al.*, 1989). The *groEL* gene product of *E. coli* belongs to a second important class of stress proteins (hsp60) which includes the mycobacterial 65 kD antigen and a family of 55–65 kD proteins present in the mitochondria and chloroplasts of higher organisms (Young *et al.*, 1988a; Hemmingsen *et al.*, 1988; Jindal *et al.*, 1989). A third group of stress proteins present in eukaryotic cells have molecular weights ranging from 80–100 kD, referred to as the hsp90 gene family. The *E. coli* protein HtpG (or 'C62.5') belongs to the hsp90 family (Bardwell and Craig, 1987) but, although a related protein can be detected in *M. tuberculosis* using appropriate antibody and oligonucleotide probes (P. Norton, R. Lathigra and D. Young, unpublished results), it has not been found amongst the mycobacterial antigens identified to date. A group of loosely related low molecular weight heat-shock proteins with structural features similar to alpha-crystallin are also found in eukaryotic cells, and the *M. leprae* 18 kD antigen is apparently unique in being a bacterial member of this family (Nerland *et al.*, 1988). Many other heat-shock proteins have been identified in *E. coli*, including the products of the *dnaJ* and *grpE* genes, which have functional interactions with *DnaK*, and the *groES* gene product which interacts with GroEL and corresponds to the 12 kD antigen of *M. tuberculosis* (Baird *et al.*, 1988; Shinnick *et al.*, 1989).

4.1.1 Immunogenicity

The finding that major antigens belong to heat-shock protein families is not unique to mycobacterial disease, but rather seems to hold true for a wide diversity of bacterial, protozoan and helminth infections (reviewed in Young *et al.*, 1989). Hsp70 proteins are immunodominant in the antibody response to schistosomiasis, leishmaniasis, and lymphatic and ocular filariasis. In malaria and Chagas' disease, immunodominant responses to several members of hsp70 and hsp90 gene families have been found, and members of the hsp60 family have been identified as major antigens involved in immune responses to Q fever, syphilis, legionaires' disease and blinding trachoma. In addition, elevated immune responses to stress proteins have also been reported in association with a variety of autoimmune disorders including reactive arthritis, rheumatoid arthritis and systemic lupus erythematosus (Minota *et al.*, 1988a, b; Res *et al.*, 1988; Lamb *et al.*, 1989; Gaston *et al.*, 1989).

The origin of the strong immunogenicity of stress protein families is not clearly understood. Stress proteins are major components of all cells, but perhaps elevated synthesis in response to physiological stress associated with intracellular infection results in enhanced availability for T-cell recognition. Alternatively, it can be proposed that stress proteins have structural or functional properties — such as their affinity for denatured polypeptides — which provide them with preferential access to antigen processing compartments and consequently preferential T-cell recognition. The localization of two hsp70 genes within the MHC region of human chromosome 6 (Sargent *et al.*, 1989) is suggestive of some role in antigen processing and presentation. The recent observations regarding recognition of the mycobacterial 65 kD antigen by gamma–delta T cells has led to a suggestion that the enhanced immunogenicity of stress proteins may be a reflection of a fundamental role for these antigens during evolution of the immune system (Young and Elliott, 1989). It is proposed that stress protein recognition by gamma–delta T cells is part of a primitive system of immune surveillance from which the more specific alpha–beta T-cell responses have subsequently evolved.

4.1.2   Function

4.1.2.1   Eukaryotic cells

Members of heat-shock protein families are major constituents of all cells, and many recent studies have attempted to identify their functional role. A general concept which can be applied to the three main protein families is that they act as 'molecular chaperones' with the ability to promote or prevent other protein–protein interactions within the cell (Ellis, 1987). For example, in eukaryotic cells hsp70 proteins mediate the translocation of newly synthesized polypeptides across intracellular membranes (Chirico *et al.*, 1988; Deshaies *et al.*, 1988), promote uncoating of clathrin baskets (Chappell *et al.*, 1986), and interact with glycosylated proteins in the endoplasmic reticulum (Pelham, 1989) and with certain tumour antigens (Pinhasi-Kimhi *et al.*, 1986). Hsp60 'chaperonins' promote assembly of multisubunit complexes including the chloroplast enzyme ribulose biphosphate carboxylase oxygenase (Rubisco) (Hemmingsen *et al.*, 1988) and mitochondrial F1 ATPase (Cheng *et al.*, 1989). Hsp90 proteins interact with a variety of proteins including steroid hormone receptors and oncogene products (Schuh *et al.*, 1985; Sanchez *et al.*, 1985) in order to mask their functional activity until, in response to an appropriate signal, dissociation of hsp90 activates the protein. The precise biochemical mechanisms involved in these activities have not been determined, but energy from ATP hydrolysis clearly plays a role in the function of hsp70 and hsp60. During stress, due to heat shock or other stimuli, it is proposed that stress proteins function by stabilization of intracellular protein structures and by disruption of aggregates formed by partially denatured polypeptides (Pelham, 1986).

4.1.2.2   E. coli

The major heat-shock proteins of *E. coli* — DnaK, GroEL, GroES, DnaJ and GrpE — were originally identified on the basis of their involvement in phage replication (Georgopoulos *et al.*, 1989). DnaK acts in combination with DnaJ and GrpE to disrupt a protein complex formed at the origin of replication of phage lambda, resulting in release of the lambda P protein and initiation of DNA replication. The GroE proteins promote assembly of the B protein of lambda into a 12-mer structure which forms the head–tail connector of the phage particle. Genetic studies have shown that the heat-shock proteins also play important roles in the metabolism of *E. coli* itself. Mutants with a deletion of the *dnaK* gene can survive at 30°C but show multiple phenotypic changes, including a defect in cell division which is reversed by overproduction of the *ftsZ* gene product (Bukau *et al.*, 1989). GroE proteins are essential for survival of *E. coli* and deletion mutants have not been isolated. The key role of *groE* in *E. coli* metabolism is illustrated by consideration of mutants with deletions in the *rpoH* gene which codes for the sigma factor involved in regulation of the heat-shock response (see below). *RpoH* deletion mutants fail to grow above 20°C, but overexpression of GroE proteins suppresses the temperature sensitivity of such mutants, allowing growth up to 37°C (Kusukawa and Yura, 1988). GroE proteins are probably involved in multiple interactions with many *E. coli* proteins. Genetic evidence indicates a role in DNA replication, through interactions with the DnaA and ssb-1 proteins (Fayet *et al.*, 1986; Jenkins *et al.*, 1986; Ruben *et al.*, 1988), and again an involvement with cell division is indicated by the dependence of *groE* expression on doubling time in *E. coli* (Neidhardt *et al.*, 1981) and evidence of a link between GroES and the LetD protein of F plasmid (Miki *et al.*, 1988). GroEL is also thought to have an important chaperone function in protein translocation in *E. coli* — similar to that described for hsp70 in higher organisms (Kusukawa *et al.*, 1989).

4.1.2.3   Mycobacteria

Heat-shock proteins are clearly involved in multiple metabolic functions in *E. coli* and, directly or indirectly, have a role in control of processes of DNA replication, cell division, and protein secretion. The rate of cell division in pathogenic mycobacteria is radically different from that in *E. coli* and it is of interest to determine whether this is reflected by differences between the two organisms in the regulation and function of heat-shock proteins.

The high degree of sequence identity between heat-shock proteins of mycobacteria and *E. coli* suggests that the mycobacterial antigens will function in an analogous manner to their *E. coli* counterparts. Attempts to complement temperature sensitive and deletion mutants of *E. coli* with the mycobacterial DnaK and GroEL homologues have been unsuccessful, however, suggesting perhaps that variable regions of the proteins are important in mediating species-specific interactions with their particular protein substrates. In *E. coli*, GroEL has a functional interaction with the product of the adjacent *groES* gene, and it is possible that the mycobacterial 65 kD antigen may similarly require the presence of its own GroES partner (the *M. tuberculosis* 12 kD antigen) for expression of functional activity. Similarly *E. coli* DnaK functions along with accessory proteins DnaJ and GrpE. An open reading frame corresponding to the *M. tuberculosis dnaJ* gene is located 788 base pairs downstream of the gene for the 71 kD antigen (Lathigra *et al.*, 1988) and we have recently used glutathione transferase fusion products to prepare antisera to the mycobacterial DnaJ protein in order to investigate its potential immunogenicity and functional interaction with DnaK (T. Garbe, unpublished results).

4.1.3 Regulation

Prior to the recent interest in the function of heat-shock proteins, the heat-shock response was used primarily as a model system for studying differential gene expression. A large amount of information has consequently been accumulated regarding promoter elements and transcription factors involved in the response in eukaryotic and bacterial cells (reviewed by Lindquist, 1986; Lindquist and Craig, 1988). Little is known at present about the regulation of gene expression in mycobacteria, and the availability of genetic and immunological probes for several of the heat-shock proteins makes the stress response an attractive model system for initiating such studies. In addition, it is possible that the stress response has a physiological role during mycobacterial infection. Expression of many of the key factors which mediate interactions between bacterial pathogens and host cells is subject to induction in response to environmental signals during infection (Finlay and Falkow, 1989). Laboratory stresses — such as heat shock and exposure to oxidative radicals — may mimic the type of environmental changes encountered by intracellular parasites and may provide clues as to patterns of gene expression *in vivo*. The response to oxidative stress plays a role in intracellular survival of *Salmonella typhimurium*, for example (Morgan *et al.*, 1986; Fields *et al.*, 1986), and important virulence determinants of *Listeria monocytogenes* can be induced by severe heat shock (Sokolovic and Gobel, 1989). Understanding the regulation of the stress response in mycobacteria might therefore contribute to understanding of mechanisms involved in mycobacterial pathogenicity.

We have investigated the heat-shock response of logarithmic cultures of *M. bovis* BCG by metabolic labelling with ^{35}S-methionine for two hours at the control and elevated temperature. A 'moderate' heat shock (30–37°C or 37–42°C) leads to induction of the 71 kD and 65 kD antigens. A 'severe' heat shock (30–45°C or 37–48°C) results in decreased synthesis of the 65 kD antigen, although the 71 kD antigen remains one of the most prominent bands. Other major heat-shock proteins which do not correspond to known antigens are found at 85 kD and 15 kD, and a band which lines up with the *M. tuberculosis* 19 kD antigen is seen following severe heat shock (Figure 1.4). The kinetics of the heat-shock response in *M. bovis* BCG are

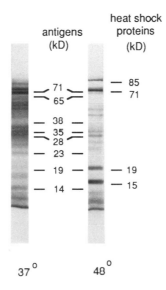

Figure 1.4   Heat shock response of *M. bovis* BCG. A logarithmic culture of *M. bovis* BCG was incubated with radiolabelled methionine for 2 h at the control temperature and with a severe heat shock. Protein synthesis was monitored by SDS-PAGE and autoradiography of nitrocellulose blots. The same blots were incubated with monoclonal antibodies in order to identify the migration position of major protein antigens. The positions of the antigens are indicated, and the molecular weights of major heat shock proteins are shown on the right of the figure.

During severe heat shock, synthesis of the 65 kD antigen is reduced, while the 71 kD antigen remains one of the most prominent bands. It is clear that the protein antigens are not all heat shock proteins, and that not all heat shock proteins are amongst the major antigens identified to date.

significantly slower than in *E. coli* — with induction taking about 10 minutes. The *M. bovis* BCG response is dependent on transcriptional induction — as judged by rifampicin sensitivity — and the slower kinetics presumably reflect a lower rate of transcription/translation in the slow growing bacteria. Under moderate heat-shock conditions, induction of the 15 kD protein increases after about one hour of incubation, suggesting some cumulative effect of chronic stress.

In *E. coli*, stress proteins are controlled in the form of regulons — unlinked genes which are co-regulated in response to a specific signal. Different stress signals induce different partially overlapping patterns of protein synthesis with particular genes, such as *dnaK*, being part of many distinct regulons (van Bogelen *et al.*, 1987). Induction of expression of the heat-shock proteins is accomplished by an increase in the concentration and activity of a particular sigma subunit of RNA polymerase (sigma 32, the *rpoH* gene product) which recognizes a heat-shock promoter in the regulatory region of the heat-shock protein genes (Straus *et al.*, 1987; Ueshima *et al.*, 1989). The increase in sigma 32 concentration is brought about by a combination of decreased proteolysis and elevated expression. Transcription of the *rpoH* gene at high temperatures requires an additional sigma factor, sigma E, which recognizes a further promoter element termed P3 (Erickson and Gross, 1989). A subset of sigma 32 independent heat-shock proteins, including the HtrA protein, is also under the control of a P3 promoter (Lipinska *et al.*, 1988).

Efficient expression of mycobacterial genes in *E. coli* generally requires the presence of an exogenous promoter — indicating that most mycobacterial promoters function only very poorly in *E. coli* (Clark-Curtiss *et al.*, 1985). The heat-shock proteins appear to be an exception to this rule in that they are readily expressed from their own promoters in a variety of vectors (e.g. Thole *et al.*, 1985). Perhaps, like the proteins themselves, the regulatory mechanisms involved in the heat-shock response are conserved between widely differing bacteria? Constructs in which transcription and translation of the beta-galactosidase gene of *E. coli* is placed under the control of regulatory elements from mycobacterial heat-shock proteins have been used to study the mycobacterial promoters (Shinnick *et al.*, 1988; R. Lathigra, unpublished results). Promoters from the 65 kD and 71 kD genes of *M. tuberculosis* are both heat inducible in *E. coli*. Although the transcriptional start sites have not yet been determined, the upstream sequences have been scanned for the presence of promoter consensus elements. The mycobacterial *groEL* and *groES* genes both contain potential sigma 32 recognition sites (Table 1.5), but such a sequence is absent from the *dnaK* gene. There is, however, a potential P3 promoter on this gene (Table 1.5), and expression of the relevant beta-galactosidase construct occurs in an *rpoH* amber mutant (R. Lathigra, unpublished results.

Table 1.5 Potential Promoter Sequences in Genes for *M. tuberculosis* Heat Shock Proteins.

	-35 region		-10 region
sigma 32 concensus	T - t C - Cc CTTGAA	13-15 bp	CCCCATt Ta
65 kD antigen	TAGCCGGGTTGCC	13 bp	CCCCGTTTC
12 kD antigen	GGGCGCCCTTGAG	9 bp	TCTCATGTA
sigma E concensus	GAACTT 4 6 bp ATAAA 4-6 bp		TCTCA 2-3bp AACA
71 kD antigen	TAAGTT 6 bp ATCAG 6 bp		TCTGA 3 bp GACA

Upstream regions of the genes coding for heat-shock proteins of *M. tuberculosis* were analysed for the presence of concensus promoter sequences present on genes regulated by sigma 32 and by sigma E in *E. coli* (Cowing *et al.*, 1985; Lipinska *et al.*, 1988). Detailed analysis of mRNA will be necessary in order to determine whether the concensus sequences shown here do in fact correspond to transcriptional start sites. Sequence data — 65 kD: Shinnick, 1987; 12 kD: Shinnick *et al.*, 1989; 71 kD: R. Lathigra (unpublished).

In *E. coli*, the *groEL* and *groES* genes form a single operon (Hemmingsen *et al.*, 1988). In mycobacteria, however, the homologous genes have separate promoters and do not appear to be closely linked (Shinnick *et al.*, 1989). Bearing in mind the above discussion on the role of the GroE proteins in cell division, it will be of interest to determine whether there are functional consequences of the dissociation between the two *groE* genes in mycobacteria.

Further studies concerning regulation of the mycobacterial stress response will be carried out along with development of gene transfer systems for mycobacteria. Reporter gene constructs, analogous to the beta-galactosidase vectors described above, can be introduced into pathogenic mycobacteria in order to address questions concerning the *in vivo* regulation of the stress response. Heat-shock promoters — which are functional in both mycobacteria and *E. coli* — will be useful in controlling expression of recombinant genes in *E. coli*–mycobacteria shuttle vectors (Snapper *et al.*, 1988).

4.2 Secreted proteins

4.2.1 Subcellular location of proteins

Several of the antigens which have been studied in detail were selected on the basis of their presence in the culture medium of *M. tuberculosis* or *M. bovis*

BCG (see Section 2.1.3). In addition, some of the antigens designated as immunodominant by monoclonal antibody analysis have also subsequently been shown to be present in the extracellular fluid (Abou-Zeid *et al.*, 1988a). As discussed above, it is possible that such secreted antigens differ from cytoplasmic antigens in a qualitative, or perhaps kinetic, manner in the way that they interact with the immune system. In addition, secreted proteins may be important in mediating the interaction between pathogenic mycobacteria and their eukaryotic host, and study of their regulation and function may provide important insights into the *in vivo* lifestyle of the pathogen.

Polypeptides which are destined for translocation across a membrane can generally be identified by the presence of a signal peptide — consisting of a few positively charged amino acids, followed by a hydrophobic stretch — preceding the N-terminus of the mature protein (see Oliver, 1985 for review). Comparison of complete nucleotide sequences with N-terminal amino acid

Table 1.6   Signal Peptide Sequences in Genes for Secreted Antigens.

Antigen	Derived amino acid sequence
α antigen (BCG85B)	M T D V S R K I R A W G R R L M I G T A A A V V L P G L V G L A G G A A T A G A F S R
P32 (BCG85A)	M Q L V D R V R G A V T G M S R R L V V G A V A R L V S G L V G A V G G T A T A G A F S R
MPB 64	M R I K I F M L V T A V V L L C C S G V A T A A P K
MPB 70	M K V K N T I A A T S F A A A G L A A L A V A V S P P A A A G D L
28 kD (b) (*M. leprae*)	M P N R R R C K L S T A I S T V A T L A I A S P C
38 kD (*M. tuberculosis*)	M K I R L H T L L A V L T A A P L L L A A A G C G S
19 kD (*M. tuberculosis*)	V K R G L T V A V A G A A I L V A G L S G C S S

Signal peptide sequences are shown using the single letter code for amino acids, with positively charged residues underlined. In each case, the three outlined residues denote the N-terminus of the mature protein — as determined experimentally (α antigen, P32, MBP64 and MPB70), or predicted by sequence analysis. The proposed N-terminus for the 38 kD and 19 kD antigens is based on the presence of a lipoprotein consensus sequence, with predicted acylation of the cysteine residue. Sequence data from: Cherayil and Young, 1988 (28 kD); Matsuo *et al.*, 1988 (α antigen); Borremans *et al.*, 1989 (P32); Yamaguchi *et al.*, 1989 (MPB64); Terasaka *et al.*, 1989 (MPB70); Andersen and Hansen, 1989 (38 kD); Ashbridge *et al.*, 1989 (19 kD).

sequences has demonstrated the presence of signal peptides associated with three of the secreted proteins of *M. bovis* BCG (the alpha antigen, MPB64 and MPB70) and the P32 secreted antigen of *M. tuberculosis* (Table 1.6). Potential signal peptides are also present in the derived sequences of the 38 kD and 19 kD antigens of *M. tuberculosis* and a 28 kD antigen of *M. leprae* (Table 1.6). The possibility of the 28 kD antigen of *M. leprae* being involved in iron transport is discussed by Dale (this volume).

In contrast to these proteins, several antigens which lack a signal peptide have also been found in mycobacterial culture media. Under appropriate conditions, for example, three of the heat-shock proteins — the 71 kD, 65 kD and 12 kD antigens — have been reported to accumulate in the culture medium (Abou-Zeid *et al.*, 1988a; de Bruyn *et al.*, 1987; Yamaguchi *et al.*, 1988). Similarly, superoxide dismutase — which corresponds to the monoclonal antibody defined 23 kD/28 kD antigen of *M. tuberculosis/M. leprae* (H. Thangaraj and M. J. Colston, personal communication) — is found in mycobacterial culture filtrates (Mayer and Falkinham, 1986; Abou-Zeid *et al.*, 1988a) even though the *M. tuberculosis* gene lacks a signal peptide (Y. Zhang, R. Lathigra, D. Young, unpublished data). Only a limited number of mycobacterial proteins are detected in the early culture filtrate (see Figure 1.3) and the mechanisms underlying the preferential accumulation of these particular antigens remain to be established.

An additional feature of some of the sequences shown in Table 1.6 is the presence of a distinctive set of residues which correspond to a lipoprotein concensus element. Several bacterial proteins are post-translationally modified by addition of fatty acid moieties to an N-terminal cysteine residue generated as a result of removal of the signal peptide (see Wu and Tokunaga, 1986 for review), and it seems likely that the 38 kD and 19 kD antigens of *M. tuberculosis* belong to this class of lipoproteins. This inference is supported by biochemical fractionation studies in which these antigens show a relative hydrophobicity in Triton X-114 phase separation protocols, and by preliminary results of palmitic acid labelling experiments (D. Young, unpublished results).

Figure 1.5 provides a schematic representation of the subcellular location of different classes of mycobacterial antigens. The heat-shock proteins and superoxide dismutase are major cytoplasmic antigens although, as noted above, the same proteins can also be found in early culture filtrates. A 31 kD antigen has been reported to associate with the cell membrane of mycobacteria belonging to the *M. avium-intracellulare* group (George and Falkinham, 1989), and the *M. leprae* membrane is characterized by high concentrations of the 35 kD antigen and a 22 kD protein. (S. W. Hunter and P. J. Brennan, personal communication). Cell wall fractions contain SDS-extractable proteins — including the 30/31 kD antigen (Barnes *et al.*, 1990) —

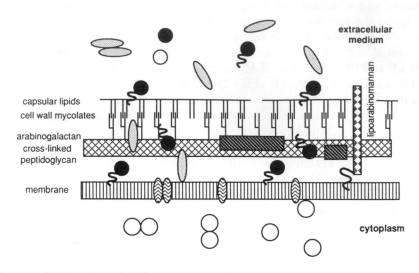

Figure 1.5   Location of different classes of protein antigens. The defined protein antigens of mycobacteria show differences in their subcellular distribution.

○ Heat shock proteins (70/71 kD, 65 kD, 18 kD, 12 kD) and superoxide dismutase (23/28 kD) are predominantly cytoplasmic proteins, although they can also be found in the supernatant of logarithmic phase cultures.

  Proteins which are found in association with the cell membrane include the 35 kD antigen of *M. leprae* and a 31 kD antigen of *M. avium-intracellulare*.

▧ A strongly immunogenic SDS-insoluble protein aggregate is tightly associated with the cell wall of *M. leprae*.

● The 38 kD and 19 kD antigens of *M. tuberculosis* have biochemical properties and sequence features which suggest that they may represent a class of lipoproteins, possibly involved in transport of nutrients through the mycobacterial cell wall. These proteins are also found in the culture medium.

◊ Major exported antigens include the BCG85 complex and the MPB64 and MPB70 antigens. These proteins are synthesized with a signal peptide and are actively secreted across the membrane. Members of the BCG85 complex are also found in a cell wall associated form.

and an SDS-insoluble high molecular weight material possibly formed by aggregation of soluble cytoplasmic proteins (Hunter *et al.*, 1989; Mehra *et al.*, 1989). It is proposed that the lipoprotein antigens are also loosely associated with the cell wall or membrane, being anchored by their lipid tail. Finally, in the extracellular fluid there is a mixture of lipoprotein and non-lipoprotein antigens exported by signal peptide mediated translocation, along with selected cytoplasmic proteins discussed above.

4.2.2 Function

Mycobacteria clearly expend considerable energy on the production of secreted proteins, and it can be assumed that they fulfill a function which is important for some aspect of bacterial growth. An exciting clue as to the possible function of at least some of the exported proteins is provided by the sequence homology identified between the 38 kD antigen of *M. tuberculosis* and the *phoS* (or *pstS*) gene product of *E. coli* (Andersen and Hansen, 1989, see also Dale, this volume, for examination of possible *pho* box in 38 kD promoter element). PhoS belongs to a class of molecules termed periplasmic binding proteins which have a well-characterized role in nutrient transport in Gram-negative bacteria (Ames, 1986). The role of the PhoS protein in phosphate transport in *E. coli* is shown in Figure 1.6 (Tommassen and Lugtenberg, 1982). Phosphate crosses the outer membrane via pores formed by the PhoE protein. PhoS then binds the phosphate and shuttles it through the periplasmic space, delivering it to a specific phosphate permease system associated with the inner membrane. Periplasmic binding proteins are readily identified by their release from Gram-negative bacteria following osmotic shock procedures which preferentially disrupt the outer membrane structure and, in addition to PhoS, binding proteins for maltose, histidine, oligopeptides, and sulphate have been extensively analysed (Ames, 1986). It has only recently been realized, however, that Gram-positive bacteria synthesize

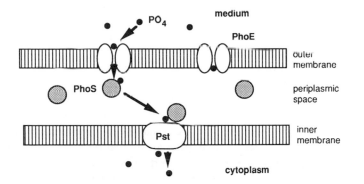

Figure 1.6   Role of periplasmic binding proteins in nutrient uptake. The function of periplasmic binding proteins in Gram negative bacteria is illustrated in the case of phosphate uptake. Phosphate ions cross the outer membrane through pores formed by the PhoE protein. The periplasmic binding protein — PhoS — then transports the phosphate to the high-affinity permease (Pst), which carries out active transport across the inner membrane.

The 38 kD antigen of *M. tuberculosis* is closely related to PhoS, and it is proposed that binding proteins may play an important role in mycobacterial nutrition.

related binding proteins (Gilson *et al.*, 1988). Gram-positive bacteria have no outer membrane and the binding proteins cannot therefore be contained within a periplasmic compartment. Sequence analysis of potential binding proteins from *Streptococcus pneumoniae* and *Mycoplasma hyorhinis* reveals the presence of lipoprotein concensus elements and it is suggested that a lipid tail may act as anchor to maintain the binding protein in close proximity to the specific permease system (Gilson *et al.*, 1988).

Mycobacteria have a highly complex cell wall with an abundance of lipid and lipopolysaccharide constituents. Models of cell-wall structure in which mycolic acids and capsular lipids interact to form a loosely defined 'outer membrane' have been proposed (see Gaylord and Brennan, 1987, for example), and understanding how nutrients are transported through such a structure is of fundamental importance in analysis of the nutrition and drug permeability of mycobacteria. It is highly probable that molecules related to the periplasmic binding proteins of Gram-negative bacteria play a major role in mycobacterial transport systems and the addition of a lipid tail to such proteins can be seen as advantageous from the point of view of interactions with cell-wall or membrane structures. It is proposed therefore that a significant fraction of the secreted proteins of mycobacteria will correspond to lipoproteins with nutrient transport functions.

The potential lipoproteins make up only a subset of the secreted proteins of mycobacteria and sequence analysis of the non-lipoprotein genes has not yet provided any homologies indicative of potential functions. The possibility that some of these also have transport functions cannot be ruled out. Perhaps these proteins leak out from some form of 'periplasmic space' generated by the mycolic acid/lipid capsule structure discussed above? Alternatively, it can be imagined that some transport proteins may be free to diffuse from one bacterial cell to another. Such altruistic behaviour, as a result of which a transport protein synthesized by one cell may end up providing nutrition for an adjacent sister cell, might be consistent with the tendency of mycobacteria to flourish as clumps or small colonies rather than as dispersed single cells.

Only one group of secreted antigens has been investigated in terms of biochemical properties which may give an indication of physiological function. This is the group of proteins referred to as the BCG85 complex (Wiker *et al.*, 1986), which is seen as a prominent doublet of molecular weights 30 kD and 31 kD in Western blot experiments (Abou-Zeid *et al.*, 1988b), and which includes the alpha antigen of *M. bovis* BCG and *M. tuberculosis* P32 for which the complete sequences have been determined (Matsuo *et al.*, 1988; Borremans *et al.*, 1989). Three closely related forms of the protein have been identified, with extensive overlap in antibody recognition and N-terminal amino acid sequence (Wiker *et al.*, 1990). Screening of recombinant DNA libraries with antisera to the BCG85 complex has resulted in isolation of several distinct genes, suggesting that the different proteins may arise from separate genes rather than as a result of post-translational modification (C. Abou-Zeid, unpublished results). These proteins have the ability to bind to fibronectin (Abou-Zeid et al., 1988b) and, in a cell-associated form, are able to mediate interaction of whole bacteria with fibronectin (Ratliff et al., 1988). Similar proteins are produced by all strains of mycobacteria and the relationship between the fibronectin-binding activity (which may be important for the pathogenic strains) and the original function of the proteins remains to be elucidated.

4.2.3 Role in host–parasite interaction

Secreted proteins are part of the interface between mycobacteria and the external environment and may be important in mediating interactions between mycobacterial pathogens and host cells during infection. Fibronectin binding proteins play a role in the virulence of several extracellular pathogens (including *Streptococcus pyogenes* and *Staphylococcus aureus*) by promoting adherence to mucosal surfaces. Attachment to fibronectin is also important in the pathogenesis of *Treponema pallidum* with a possible role in avoidance of immune recognition as well as in adherence (reviewed by Finlay and Falkow, 1989). The significance of the fibronectin-binding activity of mycobacteria is not known, although a role in modulation of macrophage uptake has been discussed (Ratliff et al., 1988).

Superoxide dismutase — the 23/28 kD antigen recognized by monoclonal antibody D2D (Table 1.1) — is released into the extracellular medium of several mycobacterial pathogens and could play a role in protecting bacteria from the toxic effects of superoxide radicals generated as a result of the host phagocytic burst (Mayer and Falkinham, 1986).

Release of secreted proteins into the phagosome, or cytoplasm, of infected cells could also have functional consequences with regard to host-cell metabolism. For example, it has been shown that release of a periplasmic binding protein from *Mycoplasma hyorhinis* can alter the invasiveness of a mouse sarcoma cell line following mycoplasma infection (Dudler et al., 1988). This may represent a rather subtle form of toxicity by which an intracellular pathogen can manipulate host-cell functions in order to provide itself with a more attractive environment. If an essential nutrient or trace element is in limited supply within a cell, for example, then competition with a specific high-affinity binding protein released by the parasite may reduce its availability for metabolism by the host cell.

4.2.4 Regulation

The proposed role of secreted proteins in nutrient transport systems of mycobacteria suggests that their synthesis might be regulated in response to growth conditions. The periplasmic binding proteins of *E. coli*, for example, are generally induced in response to depletion of their specific nutrient substrate (Ames, 1986). It is predicted, therefore, that the pattern of *in vivo* protein secretion will reflect the precise nutrient requirements and growth environment of the mycobacterial cell. Proteins which are specifically induced during intracellular growth may provide exciting new targets for immunodiagnosis since they should be recognized only by individuals with progressive infection. Construction of insertional mutations in genes coding for surface or secreted proteins using specific alkaline phosphatase-linked transposons (TnPhoA) has played an important role in identification of virulence determinants in other pathogens (Mekalanos *et al.*, 1988), and development of related systems for genetic manipulation of mycobacteria will be equally important in pursuing the analysis of function and regulation of secreted proteins.

5   CONCLUDING REMARKS

More than 20 genes for mycobacterial protein antigens have been cloned and characterized over the past few years. In addition to the importance of these defined antigens as probes for dissection of the immune response to mycobacteria, it has become clear that study of the function and regulation of the proteins themselves represents a useful approach towards analysis of the physiology and genetics of mycobacterial pathogens. It is anticipated that such an analysis will subsequently feed back into our understanding of the immune response by identifying individual components and interactions which are of particular importance in determining the many possible outcomes of the encounter between man and mycobacteria.

REFERENCES

Abou-Zeid, C., Smith, I., Grange, J. M., Ratliff, T. L., Steele, J. and Rook, G. A. W. (1988a). *J. Gen. Microbiol.* **134**, 531–8.
Abou-Zeid, C., Ratliff, T. L., Wiker, H. G., Harboe, M., Bennedsen, J. and Rook, G. A. W. (1988b). *Infect. Immun.* **56**, 3046–51.
Ames, G. F-L. (1986). *Annu. Rev. Biochem.* **55**, 397–425.
Andersen, A. B. and Hansen, E. B. (1989). *Infect. Immun.* **57**, 2481–8.
Ashbridge, K. R., Booth, R. J., Watson, J. D. and Lathigra, R. B. (1989). *Nucl. Acids Res.* **17**, 1249.

Baird, P. N., Hall, L. M. C. and Coates, A. R. M. (1988). *Nucl. Acids Res.* **16**, 9047.

Bardwell, J. C. A. and Craig, E. A. (1987). *Proc. Natl. Acad. Sci. USA* **84**, 5177–81.

Barnes, P. F., Mehra, V., Hirschfield, G. R., Fong, S-J., Abou-Zeid, C., Rook, G. A. W., Hunter, S. W., Brennan, P. J. and Modlin, R. L. (1989). *J. Immunol.* **143**, 2656–62.

Bloom, B. R. and Mehra, V. (1984). *Immunol. Rev.* **84**, 5–28.

Booth, R. J., Harris, D. P., Love, J. M. and Watson, J. D. (1988). *J. Immunol.* **140**, 597–601.

Borremans, M., de Wit, L., Volckaert, G., Ooms, J., de Bruyn, J., van Vooren, J-P., Stelandre, M., Verhofstadt, R. and Content, J. (1989). *Infect. Immun.* **57**, 3123–30.

Britton, W. J., Hellqvist, L., Basten, A. and Inglis, A. S. (1986). *J. Exp. Med.* **164**, 695–708.

Bukau, B., Donnelly, C. E. and Walker, G. C. (1989). In *Stress-Induced Proteins* (eds M. L. Pardue, J. R. Feramisco and S. Lindquist), pp. 27–36. Alan R. Liss, NY.

Chappell, T. G., Welch, W. J., Schlossman, D. M., Palter, K. B., Schlesinger, M. J. and Rothman J. E. (1986). *Cell* **45**, 3–13.

Cheng, M. Y., Hartl, F-U., Martin, J., Pollock, R. A., Kalousek, F., Neupert, W., Hallberg, E. M., Hallberg, R. L. and Horwich, A. L. (1989). *Nature* **337**, 620–5.

Cherayil, B. J. and Young, R. A. (1988). *J. Immunol.* **141**, 4370–5.

Chirico, W. J., Waters, M. G. and Blobel, G. (1988). *Nature* **332**, 805–10.

Clark-Curtiss, J. E., Jacobs, W. R., Docherty, M. A., Ritchie, L. R. and Curtiss, R. (1985). *J. Bacteriol.* **161**, 1093–102.

Closs, O., Harboe, M., Axelsen, N. H., Bunch-Christensen, K. and Magnusson, M. (1980). *Scand. J. Immunol.* **12**, 249–64.

Cohen, M. L., Mayer, L. W., Rumschlag, H. S., Yakrus, M. A., Jones, W. D. and Good, R. C. (1987). *J. Clin. Microbiol.* **25**, 1176–80.

Converse, P. J., Ottenhoff, T. H. M., Gebre, N., Ehrenberg, J. P. and Kiessling, R. (1988). *Scand. J. Immunol.* **27**, 515–25.

Cowing, D. W., Bardwell, J. C. A., Craig, E. A., Woolford, C., Hendrix, R. and Gross, C. (1985). *Proc. Natl. Acad. Sci. USA* **82**, 2679–83.

Damiani, G., Bianco, A., Beltrame, A., Vismara, D., Mezzopreti, M. F., Colizzi, V., Young, D. B. and Bloom, B. R. (1988). *Infect. Immun.* **56**, 1281–7.

de Bruyn, J., Bosmans, R., Turneer, M., Weckx, M., Nyabenda, J., van Vooren, J-P., Falmagne, P., Wiker, H. G. and Harboe, M. (1987). *Infect. Immun.* **55**, 245–52.

Deshaies, R. J., Koch, B. D., Werner-Washburne, M., Craig, E. A. and Schekman, R. (1988). *Nature* **332**, 800–5.

Dockrell, H. M., Stoker, N. G., Lee, S. P., Jackson, M., Grant, K. A., Joye, N. F., Lucas, S. B., Hasan, R., Hussain, R. and McAdam, K. P. W. J. (1989). *Infect. Immun.* **57**, 1979–83.

Dudler, R., Schmidhauser, C., Parish, R. W., Wettenhall, R. E. H. and Schmidt, T. (1988). *EMBO J.* **7**, 3963–70.

Ellis, R. J. (1987). *Nature* **328**, 378–9.

Emmrich, F., Thole, J., van Embden, J. and Kaufmann, S. H. E. (1985). *J. Exp. Med.* **163**, 1024–9.

Engers, H. D. and Workshop Participants (1985). *Infect. Immun.* **48**, 603–5.

Engers, H. D. and Workshop Participants (1986). *Infect. Immun.* **51**, 718–20.

Erickson, J. W. and Gross, C. A. (1989). *Genes Dev.* **3**, 1462–71.

Espitia, C., Cervera, I., Gonzalez, R. and Mancilla R. (1989). *Clin. Exp. Immunol.* **77**, 373–7.

Fayet, O., Louarn, J-M. and Georgopoulos, C. (1986). *Mol. Gen. Genet.* **202**, 435–45.

Fields, P. I., Swanson, R. V., Haidaris, C. G. and Heffron, F. (1986). *Proc. Natl. Acad. Sci. USA* **83**, 5189–93.

Filley, E., Abou-Zeid, C., Waters, M. and Rook, G. A. W. (1989). *Immunology* **67**, 75–80.

Finlay, B. B. and Falkow, S. (1989). *Microbiol. Rev.* **53**, 210–30.

Gaston, J. S. H., Life, P. F., Bailey, L. C. and Bacon, P. A. (1989). *J. Immunol.* **143**, 2494–500.
Garsia, R. J., Hellqvist, L., Booth, R. J., Radford, A. J., Britton, W. J., Astbury, L., Trent, R. J. and Basten, A. (1989). *Infect. Immun.* **57**, 204–12.
Gaylord, H. and Brennan, P. J. (1987). *Annu. Rev. Microbiol.* **41**, 645–75.
George, K. L. and Falkinham, J. O. (1989). *Can. J. Microbiol.* **35**, 529–34.
Georgopoulos, C., Tilly, K., Ang, D., Chandrasekhar, G. N., Fayet, O., Spence, J., Ziegelhoffer, T., Liberek, K. and Zylicz, M. (1989). In *Stress-Induced Proteins* (eds M. L. Pardue, J. R. Feramisco and S. Lindquist), pp. 37–47. Alan R. Liss, NY.
Germain, R. N. (1986). *Nature* **322**, 687–9.
Gilson, E., Alloing, G., Schmidt, T., Claverys, J-P., Dudler, R. and Hofnung, M. (1988). *EMBO J.* **7**, 3971–4.
Hahn, H. and Kaufmann, S. H. E. (1981). *Rev. Infect. Dis.* **2**, 1221–50.
Harboe, M., Nagai, S., Patarroyo, M. E., Torres, M. L., Ramirez, C. and Cruz, N. (1986). *Infect. Immun.* **52**, 293–302.
Haregewoin, A., Soman, G., Hom, R. C. and Finberg, R. W. (1989). *Nature* **340**, 309–312.
Hemmingsen, S. M., Woolford, C., van der Vies, S. M., Tilly, K., Dennis, D. T., Georgopoulos, C. P., Hendrix, R. W. and Ellis, R. J. (1988). *Nature* **333**, 330–4.
Holoschitz, J., Koning, F., Coligan, J. E., de Bruyn, J. and Strober, S. (1989). *Nature* **339**, 226–9.
Hunter, S. W., McNeil, M., Modlin, R. L., Mehra, V., Bloom, B. R. and Brennan, P. J. (1989). *J. Immunol.* **142**, 2864–72.
Huygen, K., Palfliet, K., Jurion, F., Hilgers, J., ten Berg, R., van Vooren J-P. and de Bruyn, J. (1988a). *Infect. Immun.* **56**, 3198–200.
Huygen, K., van Vooren, J-P., Turneer, M., Bosmans, R., Dierckx, P. and de Bruyn, J. (1988b). *Scand. J. Immunol.* **27**, 187–94.
Ivanyi, J. and Sharp, K. (1986). *Immunology* **59**, 329–32.
Ivanyi, J., Bothamley, G. H. and Jackett, P. S. (1988). *Br. Med. Bull.* **44**, 635–44.
Janeway, C. A., Jones, B. and Hayday, A. (1988). *Immunology Today* **9**, 73–6.
Janis, E. M., Kaufmann, S. H. E., Schwartz, R. H. and Pardoll, D. M. (1989). *Science* **244**, 713–16.
Jenkins, A. J., March, J. B., Oliver, I. R. and Masters, M. (1986). *Mol. Gen. Genet.* **202**, 446–54.
Jindal, S., Dudani, A. K., Singh, B., Harley, C. B. and Gupta, R. S. (1989). *Mol. Cell. Biol.* **9**, 2279–83.
Kaufmann, S. H. E. (1988). *Immunology Today* **9**, 168–74.
Kaufmann, S. H. E., Vath, U., Thole, J. E. R., van Embden, J. D. A. and Emmrich, F. (1987). *Eur. J. Immunol.* **17**, 351–7.
Klatser, P. R., van Rens, M. M. and Eggelte, T. A. (1984). *Clin. Exp. Immunol.* **55**, 537–44.
Koga, T., Wand-Wurttenberger, A., de Bruyn, J., Munk, M. E., Schoel, B. and Kaufmann, S. H. E. (1989). *Science* **245**, 1112–15.
Kusukawa, N. and Yura, T. (1988). *Genes Dev.* **2**, 874–82.
Kusukawa, N., Yura, T., Ueguchi, C., Akiyama, Y. and Ito, K. (1989). *EMBO J.* **8**, 3517–21.
Lamb, J. R. and Young, D. B. (1987). *Immunology* **60**, 1–5.
Lamb, J. R., Ivanyi, J., Rees, A., Young, R. A. and Young, D. B. (1986). *Lepr. Rev.* **57** (suppl. 2), 131–7.
Lamb, J. R., O'Hehir, R. E. and Young, D. B. (1988). *J. Immunol. Methods* **110**, 1–10.
Lamb, J. R., Bal, V., Mendez-Samperio, P., Mehlert, A., So, A., Rothbard, J. B., Jindal, S., Young, R. A. and Young, D. B. (1989). *International Immunology* **1**, 191–6.
Lathigra, R., Young, D. B., Sweetser, D. and Young, R. A. (1988). *Nucl. Acids Res.* **16**, 1636.
Lindquist, S. (1986). *Annu. Rev. Biochem.* **55**, 1151–91.

Lindquist, S. and Craig, E. A. (1988). *Annu. Rev. Genet.* **22**, 631–7.

Lipinska, B., Sharma, S. and Georgopoulos, C. (1988). *Nucl. Acids Res.* **16**, 10053–67.

Ljungqvist, L., Worsaae, A. and Heron, I. (1988). *Infect. Immun.* **56**, 1994–8.

Mackaness, G. B. (1969). *J. Exp. Med.* **129**, 973–86.

Matsuo, K., Yamaguchi, R., Yamazaki, A., Tasaka, H. and Yamada, T. (1988). *J. Bacteriol.* **170**, 3847–54.

Mayer, B. K. and Falkinham, J. O. (1986). *Infect. Immun.* **53**, 631–5.

Mehlert, A. and Young, D. (1989). *Mol. Microbiol.* **3**, 125–30.

Mehra, V., Sweetser, D. and Young, R. A. (1986). *Proc. Natl. Acad. Sci. USA* **83**, 7013–17.

Mehra, V., Bloom, B. R., Torigian, V. K., Mandich, D., Reichel, M., Young, S. M. M., Salgame, P., Convit, J., Hunter, S. W., McNeil, M., Brennan, P. J., Rea, T. H. and Modlin, R. L. (1989). *J. Immunol.* **142**, 2873–8.

Mekalanos, J. J., Peterson, K. M., Finn, T. and Knapp, S. (1988). *Antonie van Leeuwenhoek J. Microbiol.* **54**, 379–87.

Melancon-Kaplan, J., Hunter, S. W., McNeil, M., Stewart, C., Modlin, R. L., Rea, T. H., Convit, J., Salgame, P., Mehra, V., Bloom, B. R. and Brennan, P. J. (1988). *Proc. Natl. Acad. Sci. USA* **85**, 1917–21.

Mendez-Samperio, P., Lamb, J., Bothamley, G., Stanley, P., Ellis, C. and Ivanyi, J. (1989). *J. Immunol.* **142**, 3599–604.

Miki, T., Orita, T., Furuno, M. and Horiuchi, T. (1988). *J. Mol. Biol.* **201**, 327–38.

Minden, P., Kelleher, P. J., Freed, J. H., Nielsen, L. D., Brennan, P. J., McPherson, L. and McClatchy, J. K. (1984). *Infect. Immun.* **46**, 519–25.

Minota, S., Koyasu, S., Yahara, I. and Winfield, J. (1988a). *J. Clin. Invest.* **81**, 106–9.

Minota, S., Cameron, B., Welch, W. J. and Winfield, J. B. (1988b). *J. Exp. Med.* **168**, 1475–80.

Modlin, R. L., Pirmez, C., Hofman, F. M., Tongian, V., Uyemura, K., Rea, T. H., Bloom, B. R. and Brenner, M. B. (1989). *Nature* **339**, 544–8.

Morgan, R. W., Christman, M. F., Jacobson, F. S., Storz, G. and Ames, B. N. (1986). *Proc. Natl. Acad. Sci. USA* **83**, 8059–63.

Mueller, D. L., Jenkins, M. K. and Schwartz, R. H. (1989). *Annu. Rev. Immunol.* **7**, 445–80.

Munk, M. E., Schoel, B. and Kaufmann, S. H. E. (1988). *Eur. J. Immunol.* **18**, 1835–8.

Munk, M. E., Schoel, B., Modrow, S., Karr, R. W., Young, R. A. and Kaufmann, S. H. E. (1989). *J. Immunol.* **143**, 2844–9.

Mustafa, A. S., Gill, H. K., Nerland, A., Britton, W. J., Mehra, V., Bloom, B. R., Young, R. A. and Godal, T. (1986). *Nature* **319**, 63–6.

Mustafa, A. S., Oftung, F., Deggerdel, A., Gill, H. K., Young, R. A. and Godal, T. (1988). *J. Immunol.* **141**, 2729–33.

Neidhardt, F. C., Phillips, T. A., van Bogelen, R. A., Smith, M. W., Georgalis, Y. and Subramaniam, A. R. (1981). *J. Bacteriol.* **145**, 513–20.

Nerland, A. H., Mustafa, A. S., Sweetser, D., Godal, T. and Young, R. A. (1988). *J. Bacteriol.* **170**, 5919–21.

O'Brien, R. L., Happ, M. P., Dallas, A., Palmer, E., Kubo, R. and Born, W. K. (1989). *Cell* **57**, 667–74.

Oftung, F., Mustafa, A. S., Husson, R., Young, R. A. and Godal, T. (1987). *J. Immunol.* **138**, 927–31.

Oliver, D. (1985). *Annu. Rev. Microbiol.* **39**, 615–48.

Orme, I. M. (1987). *J. Immunol.* **138**, 293–8.

Ottenhoff, T. H. M., Klatser, P. R., Ivanyi, J., Elferink, D. G., de Wit, M. Y. L. and de Vries, R. R. P. (1986a) *Nature* **319**, 66–8.

Ottenhoff, T. H. M., Elferink, D. G., Klatser, P. R. and de Vries, R. R. P. (1986b). *Nature* **322**, 462–4.

Ottenhoff, T. H. M., Kale Ab, B., van Embden, J. D. A., Thole, J. E. R. and Kiessling, R. (1988). *J. Exp. Med.* **168**, 1947–52.

Pelham, H. R. B. (1986). *Cell* **46**, 959–61.

Pelham, H. R. B. (1989). *EMBO J.* **8**, 3171–6.

Pinhasi-Kimhi, O., Michalovitz, D., Ben-Zeev, A. and Oren, M. (1986). *Nature* **320**, 182–4.

Radford, A. J., Duffield, B. J. and Plackett, P. (1988). *Infect. Immun.* **56**, 921–5.

Ratliff, T. L., McGarr, J. A., Abou-Zeid, C., Rook, G. A. W., Stanford, J. L., Aslanzadeh, J. and Brown, E. J. (1988). *J. Gen. Microbiol.* **134**, 1307–13.

Raulet, D. H. (1989). *Annu. Rev. Immunol.* **7**, 175–207.

Rees, A., Scoging, A., Mehlert, A., Young, D. B. and Ivanyi, J. (1988). *Eur. J. Immunol.* **18**, 1881–7.

Res, P. C. M., Schaar, C. G., Breedveld, F. C., van Eden, W., van Embden, J. D. A., Cohen, I. R. and de Vries, R. R. P. (1988). *Lancet* **ii**, 478–80.

Ridley, D. S. and Jopling, W. H. (1966). *Int. J. Lepr.* **34**, 255–73.

Rook, G. A. W., Steele, J., Barnass, S., Mace, J. and Stanford, J. L. (1986). *Clin. Exp. Immunol.* **63**, 105–10.

Ruben, S. M., van den Brink-Webb, S. E., Rein, D. C. and Meyer, R. R. (1988). *Proc. Natl. Acad. Sci. USA* **85**, 3767–71.

Rumschlag, H. S., Shinnick, T. S. and Cohen, M. L. (1988). *J. Clin. Microbiol.* **26**, 2200–2.

Sanchez, E. R., Toft, D. O., Schlesinger, M. J. and Pratt, W. B. (1985). *J. Biol. Chem.* **260**, 12398–401.

Sargent, C. A., Dunham, I., Trowsdale, J. and Campbell, R. D. (1989). *Proc. Natl. Acad. Sci. USA* **86**, 1968–72.

Schuh, S., Yonemoto, W., Bauer, V. J., Riehl, R. M., Sullivan, W. P. and Toft, D. O. (1985). *J. Biol. Chem.* **260**, 14292–6.

Schwartz, R. H. (1985). *Annu. Rev. Immunol.* **3**, 237–61.

Schwartz, R. H. (1989). *Cell* **57**, 1073–81.

Shinnick, T. M. (1987). *J. Bacteriol.* **169**, 1080–8.

Shinnick, T. M., Vodkin, M. H. and Williams, J. L. (1988). *Infect. Immun.* **56**, 446–51.

Shinnick, T. M., Plikaytis, B. B., Hyche, A. D., van Landingham, R. M. and Walker, L. L. (1989). *Nucl. Acids Res.* **17**, 1254.

Snapper, S. B., Lugosi, L., Jekkel, A., Melton, R. E., Kieser, T., Bloom, B. R. and Jacobs, W. R. (1988). *Proc. Natl. Acad. Sci. USA* **85**, 6987–91.

Sokolovic, Z. and Gobel, W. (1989). *Infect. Immun.* **57**, 295–8.

Straus, D. B., Walter, W. A. and Gross, C. A. (1987). *Nature* **329**, 348–51.

Terasaka, K., Yamaguchi, R., Matsuo, K., Yamazaki, A., Nagai, S. and Yamada, T. (1989). *FEMS Microbiol. Letts.* **58**, 273–6.

Thole, J. E. R., Dauwerse, H. G., Das, P. K., Groothius, D. G., Schouls, L. M. and van Embden, J. D. A. (1985). *Infect. Immun.* **50**, 800–6.

Thole, J. E. R., Keulen, W. J., Kolk, A. H. J., Groothius, D. G., Berwald, L. G., Tiesjema, R. H. and van Embden, J. D. A. (1987). *Infect. Immun.* **55**, 1466–75.

Tommassen, J. and Lugtenberg, B. (1982). *Ann. Microbiol. (Inst. Pasteur)* **133A**, 243–9.

Ueshima, R., Fujita, N. and Ishihama, A. (1989). *Mol. Gen. Genet.* **215**, 185–9.

van Bogelen, R. A., Kelley, P. M. and Neidhardt, F. C. (1987). *J. Bacteriol.* **169**, 26–32.

Vega-Lopez, F., Stoker, N. G., Locniskar, M. F., Dockrell, H. M., Grant, K. A. and McAdam, K. P. W. J. (1988). *J. Clin. Microbiol.* **26**, 2474–9.

Wiker, H. G., Harboe, M. and Lea, T. E. (1986). *Int. Archs. Allergy Appl. Immunol.* **81**, 298–306.

Wiker, H. G., Harboe, M., Bennedsen, J. & Closs, O. (1988). *Scand. J. Immunol.* **27**, 223–39.

Wiker, H. G., Sletten, K., Nagai, S. and Harboe, M. (1990). *Infect. Immun.* **58**, 272–4.

Wu, H. C. and Tokunaga, M. (1986). *Curr. Top. Microbiol. Immunol.* **125**, 127–157.

Yamaguchi, R., Matsuo, K., Yamazaki, A., Nagai, S., Terasaka, K. and Yamada, T. (1988). *FEBS Letts.* **240**, 115–17.

Yamaguchi, R., Matsuo, K., Yamazaki, A., Abe, C., Nagai, S., Terasaka, K. and Yamada, T. (1989). *Infect. Immun.* **57**, 283–8.

Young, D., Kent, L., Rees, A., Lamb, J. and Ivanyi, J. (1986). *Infect. Immun.* **54**, 177–83.

Young, D. B., Kent, L. and Young, R. A. (1987). *Infect. Immun.* **55**, 1421–5.

Young, D., Lathigra, R., Hendrix, R., Sweetser, D. and Young, R. A. (1988a). *Proc. Natl. Acad. Sci. USA* **85**, 4267–70.

Young, D. B., Mehlert, A., Bal, V., Mendez-Samperio, P., Ivanyi, J. and Lamb, J. R. (1988b). *Antonie van Leeuwenhoek J. Microbiol.* **54**, 431–9.

Young, D., Lathigra, R. and Mehlert, A. (1989). In *Stress-Induced Proteins* (eds M. L. Pardue, J. R. Feramisco and S. Lindquist), pp. 275–85. Alan R. Liss, NY.

Young, R. A. and Elliott, T. J. (1989). *Cell* **59**, 5–8.

Young, R. A., Bloom, B. R., Grosskinsky, C. M., Ivanyi, J., Thomas, D. and Davis, R. W. (1985a). *Proc. Natl. Acad. Sci. USA* **82**, 2583–7.

Young, R. A., Mehra, V., Sweetser, D., Buchanan, T. M., Clark-Curtiss, J. E., Davis, R. W. and Bloom, B. R. (1985b). *Nature* **316**, 450–2.

2 The 65 kD antigen: molecular studies on a ubiquitous antigen

Jelle E. R. Thole and Ruurd van der Zee*

*Immunohaematology and Blood Bank, University Hospital Leiden,
PO Box 9600, 2300 RC Leiden, The Netherlands
* National Institute of Public Health and Environmental Protection,
Laboratory of Bacteriology, PO Box 1, 3720 BA Bilthoven, The
Netherlands*

1 INTRODUCTION

Since the beginning of this century the characterization of mycobacterial antigens has been a major focus of research for scientists studying mycobacteria, under the assumption that these antigens are useful tools to analyse immunopathological events caused by mycobacteria. The availability of purified and well-characterized antigens is considered to be of great importance for the study of their interaction with the immune system of the host and for the development of diagnostic reagents and vaccine components. Therefore, a variety of physicochemical techniques have been used extensively in attempts to purify individual mycobacterial antigens from culture filtrates or cell extracts. However, the complex antigenic structure of mycobacteria — mainly due to their relatively lipid-rich cell wall — has been a major drawback in these studies. As a consequence, only a relatively small number of antigens have been purified in sufficient amounts to allow their biochemical characterization and immunological evaluation. For extensive reviews of these studies see Barksdale and Kim (1977); Daniel and Janicki (1978) and Goren (1982).

In recent years developments in recombinant DNA technology have provided an alternative approach to the characterization of mycobacterial

MOLECULAR BIOLOGY OF THE MYCOBACTERIA
ISBN 0–12–483378–0

protein antigens. The successful cloning and expression of mycobacterial genes in the alternative host *Escherichia coli* have allowed the characterization and purification of a number of mycobacterial protein antigens without a cumbersome biochemical isolation procedure using mycobacterial extracts (D. Young, 1988 and this volume).

One of the first mycobacterial antigens that was identified by the recombinant DNA approach was the 65 kilodalton (kD) antigen. Thorough analysis of this antigen by molecular-biological, immunological and biochemical studies revealed a number of interesting features which have led to a growing interest in this antigen beyond the field of mycobacteriology. Firstly, the 65 kD protein appeared to be a major mycobacterial antigen. In a high proportion of individuals and animals infected or immunized with mycobacteria, antibodies and T cells reactive to the 65 kD antigen could be detected. In addition, in a proportion of healthy individuals, B-cell and T-cell responses to this antigen were found. The availability of many antibodies and T cells has facilitated the dissection of B-cell and T-cell epitopes on the 65 kD protein. Secondly, the mycobacterial antigen appeared to be a member of the highly conserved family of heat-shock proteins, and corresponding proteins were found in a wide variety of prokaryotic and eukaryotic organisms. Analysis of bacterial, plant and yeast members of this family showed that they function as chaperones in post-translational assembly of certain proteins in prokaryotes, chloroplasts and mitochondria. Thirdly, from several animal models it became clear that the 65 kD antigen plays a role in the pathogenesis of auto-immune arthritis. In one model — that of adjuvant arthritis — a T-cell epitope was identified, which is recognized by T-cells that can either induce or protect against adjuvant arthritis in rats. In this chapter we attempt to review the literature regarding the various aspects of this antigen.

2   THE 65 kD ANTIGEN

2.1   Isolation and expression of the gene

When scientists first started to use recombinant DNA technology to clone and express mycobacterial DNA, little immunological data existed on the basis of which it could be decided which antigens should be selected for detailed analysis. However, as described by D. Young *et al.* in Chapter 1 of this volume, in recent years several selection criteria have been formulated.

The first criterion is the apparent immunodominancy of certain antigens in the immune response to mycobacteria. After immunization of Balb/c mice with cell extracts and culture filtrates of *Mycobacterium tuberculosis* and

M. leprae, most of the generated monoclonal antibodies were found to be directed to a small group of proteins, suggesting that these proteins are immunodominant mycobacterial antigens (Engers *et al.*, 1985, 1986). Among these proteins, the 65 kD antigen was most frequently recognized by monoclonal antibodies. Thus, the 65 kD protein appeared to be an immunodominant protein in the B-cell response of Balb/c mice. Later studies showed that the 65 kD antigen is also a major target in the T-cell response in mice, and that in humans too, the 65 kD antigen appears to be an important B-cell and T-cell antigen in the immune response to mycobacteria (see below, Section 4).

Another criterion is the immunodiagnostic potential of certain mycobacterial antigens. Immunological analysis of the 65 kD antigen using polyclonal and monoclonal antibodies revealed that this protein is present in many other mycobacteria and that it mainly contains conserved epitopes (Thole *et al.*, 1985; Buchanan *et al.*, 1987). The immunological conservedness of the 65 kD antigen among mycobacteria was discovered earlier by crossed immunoelectrophoresis (Closs *et al.*, 1980). Using this technique, by which more than 60 different mycobacterial antigens could be identified, the 65 kD protein was found to correspond to antigen 82 of the *M. bovis* BCG reference system for mycobacterial antigens (Harboe and Wiker, 1986). Although almost exclusively conserved epitopes were present on the 65 kD protein, two monoclonal antibodies were described that recognized specific epitopes on the *M. leprae* and *M. tuberculosis* complex antigen, respectively (Coates *et al.*, 1981; Buchanan *et al.*, 1987). Characterization of these speciesspecific epitopes on this immunodominant protein might be useful for the development of immunodiagnostic reagents (Ivanyi *et al.*, 1985). Thus, both its apparent immunodominancy and its immunodiagnostic potential merit detailed analysis of the 65 kD antigen.

Very soon after recombinant DNA technology began to be applied to clone and express mycobacterial genes, it became clear that mycobacterial DNA is generally expressed very poorly in *E. coli* (Clark-Curtiss *et al.*, 1985; Jacobs *et al.*, 1986; see Dale, this volume). Therefore, expression vectors carrying *E. coli*-derived transcription and translation signals were used to construct genomic libraries of mycobacteria. The most frequently used vector to construct these libraries has been lambda gt11. This vector carries *E. coli*derived transcription and translation signals of the *lacZ* gene, and promotes expression of beta-galactosidase fusion proteins (R. Young and Davis, 1983). Using monoclonal and polyclonal antibodies recognizing the 65 kD protein from genomic libraries of *M. tuberculosis*, *M. africanum*, *M. bovis* BCG, and *M. leprae*, two groups of lambda gt11 recombinants were selected that expressed the 65 kD antigen (R. Young *et al.*, 1985a, b; Lu *et al.*, 1987; Husson and Young, 1987; Shinnick, 1987; D. Young *et al.*, 1987b; Andersen *et al.*, 1988; Table 2.1). The first group of recombinants contained part of the

Table 2.1 Expression of 65 kD Antigen in *E. coli*.

Species	Expression		Vector	References
	Native	Fusion		
M. tuberculosis	+	+	LAMBDAGT11	R. Young (1985a)
				D. Young (1987b)
				Shinnick (1987)
				Husson (1987)
				Andersen (1988)
M. bovis BCG	+	+	LAMBDAGT11/EMBL3	Thole (1985)
				Lu (1987)
M. africanum	+	–	LAMBDAGT11	Lu (1987)
M. leprae	+	+	LAMBDAGT11	R. Young (1985b)
				Husson and
				Young (1987)

65 kD gene fused to the *lacZ* gene and consequently expressed the corresponding part of the 65 kD protein fused to beta-galactosidase. The second group of recombinants carried the complete gene and expressed the native 65 kD protein. In the latter group, expression of the 65 kD protein was found to occur independently of the *lacZ* transcription and translation signals. This finding was confirmed when a genomic library of *M. bovis* BCG in the vector EMBL3 was screened for expression of mycobacterial proteins (Thole *et al.*, 1985). This vector does not carry specific expression signals to promote production of foreign genes in *E. coli*. From this library too, recombinants were selected that expressed the native 65 kD antigen. Thus, although most mycobacterial genes appear to be poorly expressed in *E. coli*, the mycobacterial sequences preceding the 65 kD gene seem to enable proper production of the 65 kD protein in this host. A possible explanation for this finding was obtained later, when the 65 kD protein was identified as a member of the groEL family of heat-shock proteins (see below, Section 3). Both the structural genes and the regulatory DNA sequences for the expression of these heat-shock proteins appear to be highly conserved in the genome of probably all prokaryotic cells (Lindquist, 1986; Lindquist and Craig, 1988; D. Young, *et al.* 1988). The similarity between the regulatory and gene-encoding DNA sequences for the corresponding 65 kD proteins in mycobacteria and in *E. coli* could explain the relatively good expression of the mycobacterial protein in *E. coli*.

Although the level of expression of the native 65 kD protein by most recombinants enabled its detection with monoclonal and polyclonal antibodies, other vectors were employed later for large-scale production of the protein. Subcloning of the 65 kD gene of *M. bovis* BCG into the pPlc236

A

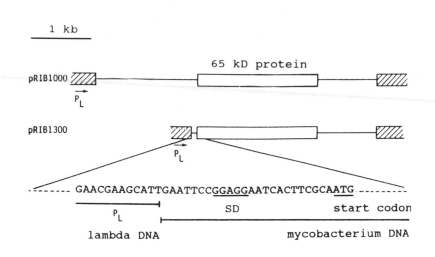

B

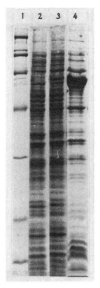

Figure 2.1 Overexpression of the *M. bovis* BCG 65 kD protein in *E. coli*. (A) Two subclones, pRIB1000 and pRIB1300, were obtained by subcloning the 65 kD gene into the pPlc236 expression vector carrying the inducible P_L promoter of bacteriophage lambda (Thole *et al.*, 1987; van Eden *et al.*, 1988). (B) SDS-PAGE of induced *E. coli* cells (lane 2), *E. coli* cells carrying pRIB1000 (lane 3), and *E. coli* cells carrying pRIB1300 (lane 4). Lane 1 contains molecular mass markers: lysozyme (14.4 kD), soybean trypsin inhibitor (21.5 kD), carbonic anhydrase (31.0 kD), ovalbumin (43.0 kD), bovine serum albumin (66.2 kD), phosphorylase *b* (92.5 kD), beta-galactosidase (116.5 kD), and myosin (200 kD). Considerable expression was found in cells carrying pRIB1000, but a much higher production was obtained when the P_L promoter was positioned just upstream of the 65 kD gene (as in pRIB1300).

vector, carrying the strong P_L promoter of bacteriophage lambda, resulted in recombinants that produced the 65 kD gene up to 20% of the total protein content after induction of the promoter (Thole *et al.*, 1987; van Eden *et al.*, 1988; Figure 2.1). Recombinants producing high amounts of the *M. leprae* protein were obtained using the pUC8 vector, which carries the inducible *lacZ* promoter (F. Lamb *et al.*, 1988). High expression of the *M. leprae* 65 kD protein was found to be due to the gene dosage effect induced by the high copy number of the pUC8 vector, rather than to induction of the *lacZ* promoter. From the soluble fraction of such overproducing *E. coli* strains the recombinant 65 kD protein could be purified by relatively simple procedures (Thole *et al.*, 1987; F. Lamb *et al.*, 1988). The availability of relatively large quantities of the purified recombinant protein now facilitates immunological studies of the 65 kD antigen.

2.2 Sequence analysis

DNA sequence analysis of the 65 kD genes of various mycobacteria revealed a very high identity between these sequences. The coding sequences of the complete genes of the closely related species *M. tuberculosis* and *M. bovis* BCG were 100% identical, and the partially determined sequences of the 65 kD genes from two other closely related species, *M. avium* and *M. paratuberculosis*, also showed complete identity, (Shinnick 1987; Thole *et al.*, 1987; Hance *et al.*, 1989). The *M. tuberculosis/M. bovis* BCG sequence revealed approximately 90% identity with those of the less closely related mycobacteria *M. leprae* and *M. fortuitum* (Mehra *et al.*, 1986; Hance *et al.*, 1989). Comparison of the DNA sequences of part of the 65 kD gene from the above mentioned species is shown in Figure 2.2. The high identity between the 65 kD genes of various less closely related species is particularly striking as their total genomic DNAs generally show a much lower overall identity. For example, hybridization studies indicated that the genome of *M. leprae* is only 20–30% identical to *M. tuberculosis* or *M. bovis* BCG at the genomic DNA level (Athwal *et al.*, 1984; see Clark-Curtiss, this volume).

High identity between the *M. tuberculosis/M. bovis* BCG sequence and the *M. leprae* sequence is also present upstream of the 65 kD gene (Shinnick *et al.*, 1987). Approximately 80% identity was found in the region comprising about 200 basepairs immediately upstream of both 65 kD genes. Further upstream of this 200-basepair region and downstream of the 65 kD gene only about 45% identity was found between *M. tuberculosis/M. bovis* BCG and *M. leprae*. The 200-basepair region immediately upstream of the 65 kD gene may represent a conserved transcription and translation control region.

Despite the high identities between the 65 kD fragments of these species, the small differences found in their DNA sequences enabled the selection of

```
                 118

                 G   I   E   K   A   V   E   K   V   T   E
M.tuberculosis   GGC ATC GAA AAG GCC GTG GAG AAG GTC ACC GAG

M.bovis BCG      --- --- --- --- --- --- --- --- --- --- ---
                                         D
M.leprae         --- --- --G --A --T --C --T --- --A --T ---

M.avium          --- --- --G --- --- --C --- --- --- --- ---

M.paratuberculosis --- --- --G --- --- --C --- --- --- --- ---

M.fortuitum      --- --- --G --- --- --C --- --- --- --- ---



                 129

                 T   L   L   K   G   A   K   E   V   E   T
M.tuberculosis   ACC CTG CTC AAG GGC GCC AAG GAG GTC GAG ACC

M.bovis BCG      --- --- --- --- --- --- --- --- --- --- ---
                                     D
M.leprae         --T --- --- --- -A- --T --- --- --- --A ---
                                     S
M.avium          --- --- --- --- TCG --- --- --- --- --- ---
                                     S
M.paratuberculosis --- --- --- --- TCG --- --- --- --- --- ---
                                     S
M.fortuitum      --G --- --G --- A-- --- --- --- --G --- ---



                 140

                 K   E   Q   I   A   A   T   A   A   I   S
M.tuberculosis   AAG GAG CAG ATT GCG GCC ACC GCA GCG ATT TCG

M.bovis BCG      --- --- --- --- --- --- --- --- --- --- ---

M.leprae         --- --A --A --- --T --- --T --- --- --- ---
                     D
M.avium          --- --C --- --C --T --- --- --G --C --C --C
                     D
M.paratuberculosis --- --C --- --C --T --- --- --G --C --C --C
                                                     G
M.fortuitum      --- --- --- --C --T --- --- --C -GT --C --C
```

Figure 2.2 DNA sequences of the 65 kD genes and their deduced amino acid sequences of various mycobacteria (Mehra *et al.*, 1986; Shinnick *et al.*, 1987; Thole *et al.*, 1987; Hance *et al.*, 1989). The DNA and deduced amino acid (one letter code) sequences for part of the *M. tuberculosis* antigen is shown completely. The numbers refer to the amino acids of the *M. tuberculosis* protein. The DNA and amino acid sequences for the other mycobacteria are only given if they differ from these *M. tuberculosis* sequences. The (–) symbol indicates identity with the *M. tuberculosis* sequences. The *M. tuberculosis* sequences are completely identical to the *M. bovis* BCG sequences, and the *M. avium* sequences are completely identical to the *M. paratuberculosis* sequences.

primers which were used to develop a polymerase chain reaction specific for
M. tuberculosis/M. bovis, *M. paratuberculosis/M. avium*, and *M. fortuitum*
(Hance, 1989; see McFadden, this volume). The presence of species-specific
DNA sequences on a conserved gene like the 65 kD gene, which is present in
many other prokaryotic and eukaryotic organisms (see below), may prove
useful not only for a rapid and sensitive identification of mycobacteria, but
also of many other organisms.

In concordance with the observed DNA homology between 65 kD genes,
the predicted amino acid sequences revealed considerable homology between
the 65 kD proteins of mycobacteria. The *M. tuberculosis/M. bovis* BCG
sequence shows a protein of 540 residues with a calculated molecular mass of
56.6 kD, and has 95% identity with the predicted *M. leprae* sequence of 541
amino acids (Shinnick *et al.*, 1987). Both sequences have no cysteine
residues, and most of the amino acid differences are present in their
C-terminal parts. The deduced amino acid sequences for part of the *M.
avium/M. paratuberculosis* and the *M. fortuitum* 65 kD antigen show 92% and
88% identity, respectively, with the corresponding part of the *M. tuberculosis/
M. bovis* BCG protein (Hance *et al.*, 1989; Figure 2.2). Thus, the mycobacterial 65 kD antigen appears to be highly conserved both at the nucleotide
and at the amino acid level.

From the primary structure of the 65 kD protein an extensive alpha-helical
secondary structure with no large hydrophobic regions can be predicted,
which would be consistent with a cytoplasmic or periplasmic localization of
the 65 kD protein in mycobacteria. However, so far various studies have
resulted in contradictory data on this issue. After sonic disruption of *M.
leprae* cells the 65 kD antigen was mainly associated with the insoluble
fraction, whereas immunoelectron microscopy of *M. bovis* BCG indicated a
predominantly cytoplasmic localization (Gillis *et al.*, 1985; Thole *et al.*,
1988a; Hunter *et al.*, 1989). During growth under zinc-limiting conditions
the 65 kD antigen is present as a major protein in culture filtrates (De Bruyn
et al., 1987, 1989). The possible involvement of the 65 kD protein in the
assembly of proteins (see below) together with the above-mentioned data
indicate that the protein may be temporarily associated with a variety of
intracellular and cell-wall structures and that it may be easily released into
the medium after damage to the cell wall.

3   THE 65 kD ANTIGEN BELONGS TO A FAMILY OF HIGHLY CONSERVED PROTEINS

Further immunological and DNA sequence data have revealed that the
mycobacterial 65 kD protein is not only conserved among mycobacteria but
also among a variety of other prokaryotic and eukaryotic organisms.

3.1 Common Antigen family of bacterial proteins

The bacterial Common Antigen was originally discovered some 15 years ago as a protein antigen of *E. coli*, that crossreacted with an antigen present in more than 50 other bacterial species as studied by gel immunoprecipitation techniques (Kaijser, 1975; Hoiby, 1975a). Since its discovery the corresponding antigens of *Pseudomonas aeruginosa*, *Treponema phagedenis* and *Legionella pnemophila* have been purified and their subunit molecular weights were found to be 60 kD (Sompolinsky *et al.*, 1980; Sand Peterson *et al.*, 1982; Pau

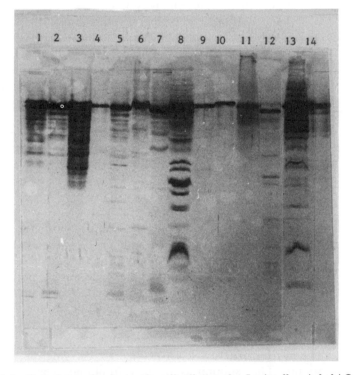

Figure 2.3 Reactivity of polyclonal antibodies to the *Legionella micdadei* Common Antigen with corresponding antigens from a variety of other bacteria (Thole *et al.*, 1988a). The Western blot was loaded with recombinant *Escherichia coli* K12 expressing the *Mycobacterium bovis* 65 kD antigen (lane 1), *E. coli* K12 (lane 2), *M. bovis* BCG, *Treponema pallidum* (lane 4), *E. coli* K12 expressing the *T. pallidum* Common Antigen (lane 5), *Klebsiella pneumoniae* (lane 6), *Yersinia enterocolitica* (lane 7), *Neisseria gonorrhoea* (lane 8), *Clostridium difficile* (lane 9), *Bacillus subtilis* (lane 10), *Nocardia asteroides* (lane 11), *Methanosarcina backeri* (lane 12), *Campylobacter jejuni* (lane 13), and *Streptococcus pyogenes* (lane 14). In all bacteria reactivity with their corresponding Common Antigen was detected.

et al., 1988). More recently, the genes encoding the Common Antigens of various bacteria were cloned and of three species, *E. coli*, *Coxiella burnetii*, and *Chlamydia psittaci* the DNA sequence was determined (Hindersson *et al.*, 1987; Hansen *et al.*, 1988; Vodkin and Williams, 1988; Hemmingsen *et al.*, 1988; Morrison *et al.*, 1989).

Immunological and DNA sequence data showed that the mycobacterial 65 kD antigen belongs to this family of bacterial proteins. Polyclonal antibodies directed to the Common Antigen of *P. aeruginosa*, *L. micdadei*, and *T. phagedenis* reacted with the 65 kD antigen of *M. tuberculosis/M. bovis* BCG in both Western blotting and crossed-immunoelectrophoresis (Thole *et al.*, 1988a). In addition, monoclonal and polyclonal antibodies directed to the mycobacterial 65 kD protein reacted with a corresponding 58–65 kD protein of a wide variety of Gram-negative bacteria, Gram-positive bacteria and archaebacteria (D. Young *et al.*, 1987b; Shinnick *et al.*, 1988; Thole *et al.*, 1988a; Figure 2.3). Crossreactivity between these proteins also occurs at the T-cell level. Human T cells derived from PPD-positive healthy individuals that responded to the mycobacterial 65 kD antigen also reacted to the corresponding *E. coli* protein (D. Young *et al.*, 1987a). The amino acid sequence inferred from DNA sequencing of the corresponding genes of *E. coli*, *C. burnetii* and *C. psittaci* revealed 55–60% identity with the *M. tuberculosis/M. bovis* 65 kD sequence (see below, Figure 2.5).

3.2 Heat-shock proteins

The Common Antigen of *E. coli*, groEL, was initially identified as a member of a family of highly conserved heat-shock proteins. Heat-shock proteins are constitutively produced in the cell, but are produced at an increased level in response to heat shock and other forms of stress (Lindquist, 1986; Lindquist and Craig, 1988). The groEL protein normally accounts for approximately 1% of the total cell protein of *E. coli* grown at 37°C, but accumulates up to 15% of total synthesis soon after the cells are shifted to 46°C (Figure 2.4). This increase in gene expression is regulated by a specific heat-shock promoter which is recognized by RNA polymerase carrying the heat shock sigma 32 subunit. This subunit is distinct from the sigma 70 subunit normally involved in *E. coli* transcription. The groEL gene is coordinately expressed with another heat-shock gene called groES, encoding for a protein of 15 kD. Corresponding bicistronic heat-shock operons were found in *C. burnetii* and in *C. psittaci* (Vodkin and Williams, 1988; Morrison *et al.*, 1989).

A sequence very similar to the *E. coli* heat-shock promoter was found upstream of the mycobacterial 65 kD gene (Chapter 1, Figure 1.5; Shinnick *et al.*, 1988). Just as in *E. coli* and other bacteria, increase in synthesis after heat shock was observed for the mycobacterial 65 kD antigen, showing that

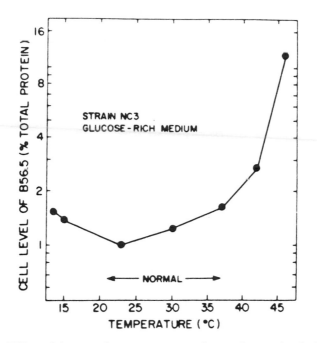

Figure 2.4 Effect of the growth temperature on the steady-state level of groEL (or B 56.5) in *Escherichia coli* (from Neidhardt *et al.*, 1981, with permission).

the mycobacterial protein, too, is a heat-shock protein (Shinnick *et al.*, 1988; D. Young *et al.*, 1988; see D. Young, this volume). In contrast to the above mentioned bacteria, in *M. tuberculosis* the corresponding 15 kD and 65 kD heat-shock genes were not localized on the same operon (Vodkin and Williams, 1988; Baird *et al.*, 1989).

The groEL family of proteins is present not only in prokaryotic but also in eukaryotic organisms. Polyclonal and monoclonal antibodies specific for the mycobacterial 65 kD protein recognized similar-sized proteins in plant, yeast and human cells (D. Young *et al.*, 1987a; Hemmingsen *et al.*, 1988; McMullin and Hallberg, 1988; van der Zee, unpublished data). At the amino acid level the mycobacterial protein was 40–50% identical with the mito-chondrial P1 protein from human HL60 cells, the hsp60 mitochondrial protein from *Saccharomyces cerevisiae*, and the plant Rubisco-binding pro-teins from *Ricinus communis* and *Triticum aestivum* (Hemmingsen *et al.*, 1988; Reading *et al.*, 1989; Jindal et al., 1989).

Comparison of the amino acid sequences of a number of prokaryotic and eukaryotic members of the groEL family is shown in Figure 2.5. Conserved regions alternating with regions of great variability in amino acid sequence

```
M.bovis BCG      MA--------------------KTIAYDEEARRGLERGLNALADAVKVTLGPKGR    36
M.tuberculosis   MA--------------------KTIAYDEEARRGLERGLNALADAVKVTLGPKGR
M.leprae         MA--------------------KTIAYDEEARRGLERGLNSLADAVKVTLGPKGR
E.coli           MA--------------------AKDVKFGNDARVKMLRGVNVLADAVKVTLGPKGR
C.burnetii       MA--------------------AKVLKFSHEVLHAMSRGVEVLANAVKVTLGPKGR
C.psittaci       MA--------------------AKNIKYNEDARKKIHKGVKTLAEAVKVTLGPKGR
S.cerevisiae     MLR-SSVVRSRATLRPLLRRAYSSH--KELKFGVEGRASLLKGVETLAEAVAATLGPKGR
human            MLRLPTVFRQMRPVSRVLAPHLTRAYAKDVKFGADARALMLQGVDLLADAVAVTMGPKGR
Ch.hamster       MLRLPTVLRQMRPVSRALAPHLTRAYAKDVKFGADARALMLQGVDLLADAVAVTMGPKGR
T.aestivum       GAD--------------------AKEIAFDQKSRAALQAGVEKLANAVGVTLGPRGR
R.communis       --------------------------RTALQSG::DKLADAVGLTLGPRGR
                                           *.. .**.**  *.**.**


M.bovis BCG      NVVLEKKWGAPTITNDGVSIAKEIELEDPYEKIGAELVKEVAKKTDDVAGDGTTTATVLA    96
M.tuberculosis   NVVLEKKWGAPTITNDGVSIAKEIELEDPYEKIGAELVKEVAKKTDDVAGDGTTTATVLA
M.leprae         NVVLEKKWGAPTITNDGVSIAKEIELEDPYEKIGAELVKEVAKKTDDVAGDGTTTATVLA
E.coli           NVVLDKSFGAPTITKDGVSVAREIELEDKFENMGAQMVKMLRGVNVLADAVKVTLGPKGR
C.burnetii       NVVLDKSFGAPTITKDGVSVAKEIELEDKFENMGAQMVKEVASRTSDDAGDGTTTATVLA
C.psittaci       HVVIDKSFGSPQVTKDGVTVAKEIELEDKHENMGAQMVKEVASKTADKAGDGTTTATVLA
S.cerevisiae     NVLIEQPFGPPKITKDGVTVAKSIVLKDKFENMGAKLLQEVASKTNEAAGDGTTSATVLG
human            TVIIEQSWGSPKVTKDGVTVAKSIDLKDKYKNIGAKLVQDVANNTNEEAGDGTTTATVLA
Ch.hamster       TVIIEQSWGSPKVTKDGVTVAKAIDLKDKYKNIGAKLVQDVANNTNEEAGDGTTTATVLA
T.aestivum       NVVLDE-YGNPKVVNDGVTIARAIELANPMENAGAALIREVASKTNDSAGDGTTTACVLA
R.communis       NVVLDE-FGSPKVVNEGVTIARAIELPDPMENAGAALIREVASKTNDSAGDGTTTASVLA
                 *....  .** ....**..*..* .  .... **  ...**.... .******.* **.


M.bovis BCG      QALVREGLRNVAAGANPLGLKRGIEKAVEKVTETLLKGAKEVETKEQIAATAAISA-GDQ    164
M.tuberculosis   QALVREGLRNVAAGANPLGLKRGIEKAVEKVTETLLKGAKEVETKEQIAATAAISA-GDQ
M.leprae         QALVKEGLRNVAAGANPLGLKRGIEKAVDKVTETLLKDAKEVETKEQIAATAAISA-GDQ
E.coli           QAIITEGLKAVAAGMNPMDLKRGIDKAVTAAVEELKALSVPCSDSKAIAQVGTISANSDE
C.burnetii       QAILVEGIKAVIAGMNPMDLKRGIDKAVTAAVAELKKISKPCKDQKAIAQVGTISANSDK
C.psittaci       EAIYSEGLRNVTAGANPMDLKRGIDKAVKVVVDEIKKISKPVQHHKEIAQVATISANNDA
S.cerevisiae     RAIFTESVKNVAAGCNPMDLRRGSQVAVEKVIEFLSANKKEITTSEEIAQVATISANGDS
human            RSIAKEGFEKISKGANPVEIRRGVMLAVDAVIAELKKQSKPVTTPEEIAQVATISANGDK
Ch.hamster       RSIAKEGFEKISKGANPVEIRRGVMLAVDAVIAELKKQSKPVTTPEEIAQVATISANGDK
T.aestivum       REIIKLGILSVTSGANPVSLKKGIDKTVQGLIEELERKARPVKGSGDIKAVASISAGNDE
R.communis       REIIKLGLLSVTSGANPVSIKRGIDKTVQGLIEELEKKARPVKGRDDIKAVASISAGNDE
                  .. ..     * **.....* .*     ...      .* ....*** .*


M.bovis BCG      SIGDLIAEAMDKVGNEGVITVEESNTFGLQLELTEGMRFDKGYISGYFVTDPERQEAVLE    214
M.tuberculosis   SIGDLIAEAMDKVGNEGVITVEESNTFGLQLELTEGMRFDKGYISGYFVTDPERQEAVLE
M.leprae         SIGDLIAEAMDKVGNEGVITVEESNTFGLQLELTEGMRFDKGYISGYFVTDAERQEAVLE
E.coli           TVGKLIAEAMDKVGKEGVITVEDGTGLQDELDVVEGMQFDRGYLSPYFINKPETGAVELE
C.burnetii       SIGDIIAEAMEKVGKEGVITVEDGSGLENALEVVEGHQFDRGYLSPYFINNQQNMSAELE
C.psittaci       EIGNLIAEAMEKVGKNGSITVEEAKGFETVLDVVEGMNFNRGYLSSYFSTNPETQECVLE
S.cerevisiae     HVGKLLASAMEKVGKEGVITIREGRTLEDELEVVEGMRFDRGFISPYFITDPKSSKVEFE
human            EIGNIISDAMKKVGRKGVITVKDGKTLNDELEIIEGMKFDRGYISPYFINTSKGQKCEFQ
Ch.hamster       DIGNIISDAMKKVGRKGVITVKDGKTLNDELEIIEGMKFDRGYISPYFINTSKGQKCEFQ
T.aestivum       LIGAMIADAIDKVGPDGVLSIESSSSFETTVDVEEGMEIDRGYISPQFVTNLEKSIVEFE
R.communis       LIGTMIADAIDKVGPDGVLSIESSSSFETTVEVEEGMEIDRGYISPQFVTNPEKLICEFE
                 .* ....*..*** .*        ...    ***...*..* *...   .*


M.bovis BCG      DPYILLVSSKVSTVKDLLPLLEKVIGAGKPLLIIAEDVEGEALSTLVVNKIRGTFKSVAV    274
M.tuberculosis   DPYILLVSSKVSTVKDLLPLLEKVIGAGKPLLIIAEDVEGEALSTLVVNKIRGTFKSVAV
M.leprae         EPYILLVSSKVSTVKDLLPLLEKVIQAGKSLLIIAEDVEGEALSTLVVNKIRGTFKSVAV
E.coli           SPFILLADKKISNIREMLPVLEAVAKAGKPLLIIAEDVEGEALATAVVNTIRGIVKVAAV
C.burnetii       NPFILLVDKKISNIRELIPLLENVAKSGRPLLVIAEDIEGEALATLVVNNIRGVVKVAAV
C.psittaci       EALVLIYDKKISGIKDFLPVLQQVAESGRPLLIIAEDIEGEALATLVVNRLRAGFRVCAV
S.cerevisiae     KPLLLLSEKKISSIQDILPALEISNQSRRPLLIIAEDVDGEALAACILNKLRGQVKVCAV
human            DAYVLLSEKKISSIQSIVPALEIANAHRKPLVIIAEDVDGEALSTLVLNRLKVGLQVVAV
Ch.hamster       DAYVLLSEKKISSVQSIVPALEIANAHRKPLVIIAEDVDGEALSTLVLNRLKVGLQVVAV
T.aestivum       NARVLITDQKITSIKEIIPLLEQTTQLRCPLFIVAEDITGEALATLVVNKLRGIINVAAI
R.communis       NARVLVTDQKITAIKDIIPLLEKTTQLRAPLLIIAEDVTGEALATLVVNKMRGILNVAAI
                 ...*..* .*........* *.    .* ..***..****...*..*  .  . *.
```

Figure 2.5 Alignment of currently known amino acid sequences of prokaryotic and eukaryotic members of the groEL family of heat-shock proteins. The sequences were aligned with the CLUSTAL programs (Higgins and Sharp, 1988, 1989) using gap penalties of 20. Below the sequences identical residues across all proteins (*) and conserved substitutions (.) are indicated.

```
M.bovis BCG    KAPGFGDRRKAMLQDMAILTGGQVISEE-VGLTLENADLSLLGKARKVVVTKDETTIVEG    333
M.tuberculosis KAPGFGDRRKAMLQDMAILTGGQVISEE-VGLTLENADLSLLGKARKVVVTKDETTIVEG
M.leprae       KAPGFGDRRKAMLQDMAILTGAQVISEE-VGLTLENTDLSLLGKARKVVMTKDETTIVEG
E.coli         KAPGFGDRRKAMLQDIATLTGGTVISEE-IGMELEKATLEDLGQAKRVVINKDTTTIIDG
C.burnetii     KAPGFGDRRKAMLQDIAVLTGGKVISEE-VGLSLEAASLDDLGSAKRVVVTKDDTTIIDG
C.psittaci     KAPGFGDRRKAMLEDIAILTGGQLISEE-LGMKLENTTLAMLGKAKKVIVSKFDTTIVEG
S.cerevisiae   KAPGFGDNRKNTIGDIAVLTGGTVFTEE-LDLKPEQCTIENLGSCDSITVTKEDTVILNG
human          KAPGFGDNRKNQLKDMAIATGGAVFGEEGLTLNLEDVQPHDLGKVGEVIVTKDDAMLLKG
Ch.hamster     KAPGFGDNRKNQLKDMAIATGGAVFGEEGLNLNLEDVQAHDLGKVGEVIVTKDDAMLLKG
T.aestivum     KAPSFGERRKAVLQDIAIVTGAEYLAKD-LGLLVENATVDQLGTARKITIHQTTTTLIAD
R.communis     KAPGFGERRKALLQDIAILTGAEFQASD-LGLLVENTSVEQLGIARKVTITKDSTTLIAD
               ***.**..**. .*.*. **.      .. ...... *    **
```

```
M.bovis BCG    AGDTDAIAGRVAQIRQEIENSDSD-YDREKLQERLAKLAGGVAVIKAGAATEVELKERKH    392
M.tuberculosis AGDTDAIAGRVAQIRQEIENSDSD-YDREKLQERLAKLAGGVAVIKAGAATEVELKERKH
M.leprae       AGDTDAIAGRVAQIRTEIENSDSD-YDREKLQERLAKLAGGVAVIKAGAATEVELKERKH
E.coli         VGEEAAIAGRVAQIRQOIEEATSD-YDREKLQERVAKLAGGVAVIKVGAATEVEMKEKKA
C.burnetii     SGDAGDIKNRVEQIRKEIENSSSD-YDREKLQERLAKLAGGVAVIKVGAATEVEMKEKKA
C.psittaci     LGGKEDIEERCESIKKQIEDSTSD-YDREKLQERLAKLSGGVAVIRVGAATEIEMKEKKD
S.cerevisiae   SGPKEAIQERIEQIKGSIDITTTNSYEKEKLQERLAKLSGGVAVIRVGGASEVEVGEKKD
human          KGDKAQIEKRIQEIIEQLDVTTSE-YEKEKLNERLAKLSDGVAVLKVGGTSDVEVNEKKD
Ch.hamster     KGEKAQIEKRIQEITEQLEITTSE-YEKEKLNERLAKLSDGVAVLKVGGTSDVEVNEKKD
T.aestivum     AASKDEIQARVAQLKKELSETDS-IYDSEKLAERIAKLSGGVAVIKVGATTETELEDRQL
R.communis     AASKDELQARIAQLKRELAETDS-VYDSEKLAERIAKLSGGVAVIKVGAATETELEDRKL
               . ...  *   ..     *,.***.**.***..****...*....*.  ...
```

```
M.bovis BCG    RIEDAVRNAKAAVEEGIVAGGGVTLLQAAPTLDELKL---EGDEATGANIVKVALEAPLK    449
M.tuberculosis RIEDAVRNAKAAVEEGIVAGGGVTLLQAAPTLDELKL---EGDEATGANIVKVALEAPLK
M.leprae       RIEDAVRNAKAAVEEGVVAGGGVTILQAAPALDKLKL---TGDEATGANIVKVALEAPLK
E.coli         RVEDALHATRAAVEEGVVAGGGVALIRVASKLADLRG--QNEDQNVGIKVALRAMEAPLR
C.burnetii     RVEDALHATRAAVEEGVVPGGGVALIRVLKSLDSVEV--ENEDQRVGVEIARRAMAYPLS
C.psittaci     RVDDAQHATLAAVEEGILPGGGTALVRCIPTLEAFIPILTNEDEQIGARIVLKALSAPLK
S.cerevisiae   RYDDALNATRAAVEEGILPGGGTALVKASRVLDEVVV--DNFDQKLGVDIIRKAITRPAK
human          RVTDALNATRAAVEEGIVLGGGCALLRCIPALDSLTP--ANEDQKIGIEIIKRTLKIPAM
Ch.hamster     RVTDALNATRAAVEEGIVLGGGCALLRCIPALDSLKP--SNEDQKIGIEIIKRALKIPAM
T.aestivum     RIEDAKNATFAAIEEGIVPGGGAAYVHLSTYVPAIKETIEDHDERLGADIIQKALQAPAS
R.communis     RIEDAKNATFAAIEEGIVPGGGAALVHLSTVVPAINGEDKDADERLGADILQKALVAPAS
               *  .**  ...  **.***,. ***    ..       .*.  *    .   *
```

```
M.bovis BCG    QIAFNSGLEPGVVAEKVRNLPAGH---GLNAQTGVYEDLLAAGVADPVKVTRSALQNAAS    506
M.tuberculosis QIAFNSGLEPGVVAEKVRNLPAGH---GLNAQTGVYEDLLAAGVADPVKVTRSALQNAAS
M.leprae       QIAFNSGMEPGVVAEKVRNLSVCH---GLNAATGEYEDLLKAGVADPVKVTRSALQNAAS
E.coli         QIVLNCGEEPSVVANTV-KGCDGNY--GYNAATEEYGNMIDMGILDPTKVTRSALQYAAS
C.burnetii     QIVKNTGVQAAVVAADKVLNHKDVNY-GYNAATGEYGDMIEMGILDPTKVTRTALQNAAS
C.psittaci     QIAANAGKEGAIICQQVLSRSSSE---GYDALRDAYTDMIEAGILDPTKVTRCALESAAS
S.cerevisiae   QIIENAGEEGSVIIGKLIDEYCDDFAKGYDASKSEYTDMLATGIIDPFKVVRSGLVDASG
human          TIAKNAGVEGSLIVEKIMQSSSE----VGYDAMAGDFVNMVEKGIIDPTKVVRTALLDAAG
Ch.hamster     TIAKNAGVEGSLIVEKILQSSSE----IGYDAMLGDFVNMVEKGIIDPTKVVRTALLDAAG
T.aestivum     LIANNAGVEGEVVIEKIKES---EWEMGYNAMTDKYENLIESGVIDPAKVTRCALQNAAS
R.communis     LIAQNAGIEGEVVVEKVKAR---EWEIGYNAMTDKYENLVEAGVIDPAKVTRCALQN---
               * .**  ...    . **.*     . .** **.*  .*
```

```
M.bovis BCG    IAGLFLTTEAVVADKPEKEKASVPG-G--GDMGGGMD-----F                   540
M.tuberculosis IAGLFLTTEAVVADKPEKEKASVPG-G--GDMGGGMD-----F
M.leprae       IAGLFLTTEAVVADKPEKTAAPASDPT--GGMGGGMD-----F
E.coli         VAGLMITTECMVTDLPKNDAADL---GAAGGMGGMGGMGGMM
C.burnetii     IAGLMITTECMVTEAPKKKEESMPGGGDMGGMGGMGGMGGMM
C.psittaci     VAGLLLTTEALIADIPEEKSSSAP--------AMPGAGMDY
S.cerevisiae   VASLLATTEVAIVDAPEPPAA--AGAGGMPGGMPGMP-G-MM
human          VASLLTTAEVVVTEIPKEEKD--PGMGAMGGMGGGMG-GGMF
Ch.hamster     VASLLTTAEAVVTEIPKEEKD--PGMGAMGGMGGGMG-GGMF
T.aestivum     VSGMVLTTQAIVVEKPKPKPKVAEPAEGQLSV----------
R.communis     ------------------------------------------
```

Figure 2.5 *Cont.*

are distributed throughout the whole length of the proteins. At the carboxyl terminus a repeated Gly-Gly-Met motif is conserved in all proteins, except for the plant Rubisco-binding proteins. In concordance with the similarity in primary structure, the predicted secondary structures of these proteins show a similar pattern of alpha helices interrupted by beta sheets.

3.3 Chaperonine function

The conservation of the groEL family throughout nature suggests an essential function for this protein in the cell. Recently, through studies of both prokaryotic and eukaryotic members, it has become clear that they have a similar structure that transiently binds to various newly synthesized polypeptides. This transient association assists in post-translational assembly

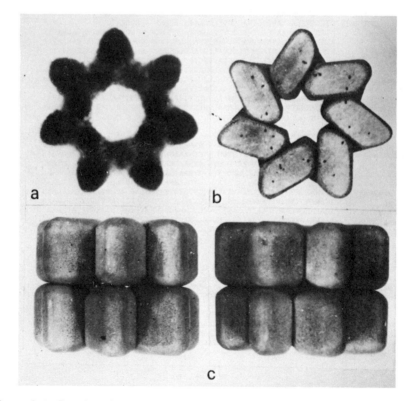

Figure 2.6 Resultant image of photosuperposition of high molecular weight particles formed by pea leaf homologue of the groEL protein of *Escherichia coli* (a) and model images of the molecule (b, c) (From Puskin *et al.*, 1982, with permission).

of these newly synthesized polypeptides into their oligomeric form, or in keeping them in the proper folding for membrane transport to other compartments. Because they are not part of the final structure of these proteins they have been called chaperonins (Ellis, 1988).

The groEL protein consists of two stacked rings of seven subunits, with a total apparent molecular mass of 700 000 (Hendrix, 1979; Hohn *et al.*, 1979; Figure 2.6). Together with the groES protein, groEL forms a complex which requires the presence of Mg-ATP (Chandrasekhar *et al.*, 1986). This complex is believed to act as the 'workbench' on which proper folding and oligomer-ization of other proteins is carried out. Both groEL and groES were shown to be essential for the correct assembly of head and tail proteins of certain bacteriophages in *E. coli* (Georgopoulos *et al.*, 1973; Sternberg, 1973; Zweig and Cummings, 1973; Kochan and Murialdo, 1983). In addition, transient binding of groEL to newly synthesized pre-beta-lactamase was shown to be essential for the secretion of beta-lactamase (Bochkareva *et al.*, 1988). Involvement in the regulation of the stability of messenger RNA and in chromosomal DNA replication were also described as possible functions for groEL, but it is not known whether these are related to its chaperonine function (Chanda *et al.*, 1985; Jenkins *et al.*, 1986; Fayet *et al.*, 1986).

A similar structure and function was described for the Rubisco-binding proteins in chloroplasts of plants and of photosynthetic prokaryotes, and for the hsp60 protein in the mitochondria of yeast cells (Miziorko and Lorimer, 1983; Ellis, 1988; Cheng *et al.*, 1989; Goloubinoff *et al.*, 1989). The Rubisco-binding protein ensures proper assembly of both the Rubisco small and large subunits into the Rubisco holoenzyme in chloroplasts, whereas the hsp60 protein was shown to be required for the proper assembly of several mitochondrial proteins. The conserved function of these proteins was further confirmed when it was shown that in recombinant *E. coli* strains carrying cDNAs which express the Rubisco small and large subunit genes from the prokaryotes *Anacystis nidulans* or *Rhodospirillum rubrum*, proper assembly of the expressed subunits into the Rubisco holoenzyme was obtained by the groEL and groES proteins of *E. coli* (Goloubinoff *et al.*, 1989). Figure 2.7 shows a model proposing the role of the Rubisco-binding protein in the assembly of Rubisco in plants.

The function of 65 kD protein of mycobacteria has not yet been studied in detail. However, the high apparent molecular weight (> 240 kD) found for the purified protein indicates that this protein, like groEL, exists in an oligomeric form (De Bruyn *et al.*, 1987; Shinnick *et al.*, 1988). Together with the immunological and amino acid similarities to other members of the groEL family of proteins, this suggests that the mycobacterial 65 kD protein, too, may function as a chaperonin.

As described above, all of the 65 kD heat-shock proteins are produced at

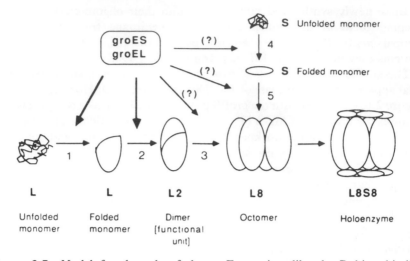

Figure 2.7 Model for the role of the groE proteins, like the Rubisco-binding protein, in the assembly of the Rubisco holoenzyme as proposed by Goloubinoff *et al.* (1989). Assembly requires the folding and multimerization of eight large and eight small subunits. In their study of the assembly of the recombinant Rubisco proteins the authors demonstrate the involvement of the groE proteins in at least folding and dimer formation of the large subunit (steps 1 and 2). Additional involvement in steps 3 and 4 was not excluded. Reproduced from Goloubinoff *et al* (1989) *Nature* **337**, 44–7, with permission. © 1989 Macmillan Magazine.

an increased level in response to heat shock. Because they have the ability to bind to a variety of polypeptides, this increased production of the groEL family of proteins may reflect a mechanism for maintaining a functional conformation of essential proteins during or following heat stress. In addition, they might assist in refolding heat-denatured proteins.

4 IMMUNOLOGICAL ANALYSIS OF THE 65 kD ANTIGEN

4.1 Immunodominance

Analysis of antibody and T-cell responses to the mycobacterial 65 kD protein indicated that this protein is a major antigen in the immune response to mycobacteria. As already mentioned, a high proportion of the monoclonal antibodies generated in Balb/c mice after immunization with *M. tuberculosis* and *M. leprae* was found to react with the 65 kD antigen (Engers *et al.*, 1985, 1986). Analysis of the T-cell response in mice immunized with *M. tuberculosis* showed that frequently T cells were encountered reacting with the 65 kD protein (Boom *et al.*, 1987; Kaufmann *et al.*, 1987). Limiting dilution analysis of uncloned T cells from C57BL/6 mice showed that approximately 10–20% of the *M. tuberculosis*-reactive T cells responded to the *M. tuberculosis*/*M. bovis* BCG 65 kD antigen.

In man, in 60–70% of sera taken from tuberculosis patients and from BCG-vaccinated individuals, a significant antibody response to the recombinant *M. tuberculosis*/*M. bovis* BCG 65 kD antigen was detected (Thole *et al.*, 1987). Also at the T cell level, reactivity to the 65 kD antigen was frequently found in tuberculosis and leprosy patients and vaccinated individuals (Emmrich *et al.*, 1986; J. Lamb *et al.*, 1986; Mustafa *et al.*, 1986; Oftung *et al.*, 1987; Ottenhoff *et al.*, 1988; van Schooten *et al.*, 1988; Ab *et al.*, 1990). In some individuals a very high proportion of the isolated T cells responded to the 65 kD antigen.

The immunodominance of the 65 kD antigen in the immune response to mycobacteria could be explained by its predominance in the mycobacterial cell. Increased production of the mycobacterial 65 kD antigen, like other members of the groEL family, was found not only in response to heat, but also when mycobacteria were grown under another stress stimulus: *M. bovis* BCG and *M. tuberculosis* cultures grown in zinc-deficient medium are significantly enriched in this protein (De Bruyn *et al.*, 1987, 1989). It might thus be expected that increased synthesis of the 65 kD protein will also occur in response to stress stimuli that mycobacteria encounter in the intracellullar environment of the host macrophages (e.g. oxidative killing mechanisms, nutrient deprivation). Because of their increased synthesis, stress-induced antigens like the 65 kD protein are thought to be important targets in the immune response against *M. tuberculosis* and *M. leprae* (D. Young *et al.*, 1988 and this volume). The possible role(s) of stress proteins in the immune response in general have been subject of several recent reviews (R. Young and Elliot, 1989; R. Young, 1990).

Another explanation for the strong immune response to the 65 kD antigen comes from studies analysing the response in healthy individuals. A reasonable proportion of healthy individuals not known to be previously exposed to mycobacteria, were shown to have antibodies and T cells reactive to the mycobacterial 65 kD antigen (Thole *et al.*, 1987; Munk *et al.*, 1988; Lamb *et al.*, 1989, Munk *et al.*, 1989). Epitope analysis revealed that these T cells were predominantly directed to conserved epitopes of the 65 kD protein present on the homologue of other bacteria and even on the human 65 kD protein (Lamb *et al.*, 1989; Munk *et al.*, 1989). Similarly, in healthy individuals antibodies to the corresponding 65 kD proteins of *P. aeruginosa* and *B. burgdorferi* have been detected (Hoiby, 1975b; Wilske *et al.*, 1986). This response in healthy humans might be due to a continuous priming of the immune system by subsequent infections with a variety of microorganisms, all expressing a similar antigenic 65 kD protein. In addition, it was suggested that this response may represent re-stimulation of immunological memory to the human 65 kD protein generated early in life. In view of this hypothesis it was recently shown that in mice a high proportion of neonatal thymocytes were shown to respond to the 65 kD antigen (O'Brien *et al.*, 1989). Whatever the origin of this immune response in healthy humans may be, these findings indicate that the 65 kD protein is not only a major target in the response to mycobacteria, but also takes part in the response to other organisms or even to autologous cells (see also below).

4.2   T-cell subsets

Although antibodies are often induced during infection, the cellular immune response plays a dominant role in the immune response to mycobacteria. T cells, by activating macrophages and by expressing cytotoxic activity, are thought to protect from mycobacterial disease, but both activities may also lead to immunopathological damage to the tissues surrounding the site of infection. In addition, the T-cell responses to mycobacteria have been implicated in the development of autoimmune diseases. The outcome of the immune response is probably dependent on the activities of a variety of T cells with different functions, which might be triggered by the multiple mycobacterial antigens.

Various subsets of T cells showing distinct functions and responding to the 65 kD antigen have been isolated from patients and healthy individuals. Most of them were characterized as antigen-specific CD4[+], CD8[−] T lymphocytes, which are likely to be helper T cells involved in triggering antibody production or macrophage activation (Emmrich *et al.*, 1986; Lamb *et al.*, 1987; Mustafa *et al.*, 1986; van Schooten *et al.*, 1988; Lamb *et al.*, 1989). Some of them, however, revealed cytotoxic activity and were able to lyse human monocytes (Oftung *et al.*, 1988; Ottenhoff *et al.*, 1988; Ab *et al.*, 1990; Munk *et al.*, 1989). A number of these cytotoxic T cells, isolated from healthy individuals, were found to respond to conserved epitopes also present on the human homologue, and thus represent potential autoreactive T cells.

More recently, CD4[−], CD8[−] T cells carrying the gamma/delta T-cell receptor and that reacted to the 65 kD protein were described (Holoshitz *et al.*, 1989; O'Brien *et al.*, 1989; Haregewoin *et al.*, 1989). These gamma/delta T cells represent a minor subset of the mature T-cell population, and their function is not well understood. It was suggested that they represent a primitive arm of the immune system devoted to eliminate stressed autologous cells (Raulet, 1989). It could be that the human 65 kD protein is an immunodominant antigen on such stressed cells. Therefore, a proportion of the gamma/delta T-cell response could be directed to the human protein.

Because of the high similarity to the human analogue such T cells also respond to the mycobacterial 65 kD protein, and might even provide immunity to mycobacteria. Because these gamma/delta T cells recognize self-epitopes on autologous cells, they may represent, like some of the cytotoxic T cells described above, T cells that could induce autoimmune disease. Although many T cells reactive to the 65 kD antigen with a variety of distinct functions have now been described, it is far from clear how these various subsets express their functions *in vivo*.

Recently, to test the protective potential of the 65 kD antigen *in vivo*, mice have been immunized with the purified protein or with recombinant live attenuated vaccines, expressing the mycobacterial antigen (see Colston, this volume; Watson, 1989). However, no significant protection against *M. leprae* and *M. tuberculosis* was detected after any of these immunizations. The role of the 65 kD antigen in autoimmunity as studied in various animal models is described below (Section 5).

The availability of a high number of antibodies and T cells recognizing specific and common antigenic sites on this well-characterized and highly conserved protein made the 65 kD antigen an attractive molecule to define the various B-cell and T-cell epitopes. Furthermore, determination of the HLA phenotype of the T cells has provided data on the relationship between the HLA-phenotypes and the T-cell epitopes that were recognized by these T cells.

4.3 B-cell epitopes

To define the antigenic sites for a number of monoclonal antibodies on the 65 kD antigen, a panel of recombinant fusion proteins containing defined parts of the *M. tuberculosis/M. bovis* BCG and the *M. leprae* 65 kD protein and, consecutively, a panel of synthetic peptides have been used. The reactivity patterns of 28 monoclonal antibodies (of 30 tested) to the recombinant fusion proteins allowed mapping of their epitopes to regions varying from 13 to 101 amino acids (Mehra *et al.*, 1986; Thole *et al.*, 1988b; see Figure 2.8). Thus, only two of the antibodies tested could not be defined by their reactivity to these fusion proteins. This suggested that most of monoclonal antibodies directed to the 65 kD antigen recognize linear determinants. Alternatively, it could indicate that some of the conformational determinants present on the native antigen are also assembled when the protein is fused to beta-galactosidase. Based on the predicted amino acid sequence of the 65 kD protein, peptides comprising the epitope-containing regions were synthesized, and by their reactivity to these synthetic peptides the antigenic sites of 10 monoclonal antibodies could be more precisely defined to regions of 22 amino acid residues or less (Brown *et al.*, 1986;

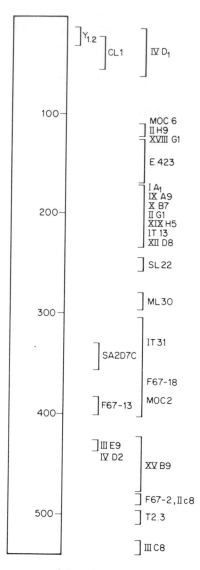

Figure 2.8   B-cell epitope map of the 65 kD antigen. The box represents the 65 kD protein and the amino acid numbers are given from top to bottom. The various regions on the 65 kD molecule containing one or more binding sites for the monoclonal antibodies are indicated to the right of the box. Most epitopes were mapped by reactivity of monoclonal antibodies to recombinant fusionproteins, however precise mapping of a number of them was further obtained by reactivity to synthetic peptides (data from Mehra *et al.*, 1986; Young *et al.*, 1987a, Thole *et al.*, 1988b, Anderson *et al.*, 1988).

Anderson *et al.*, 1988). Exact definition of essential amino acid residues for binding of five of these antibodies (IIC8, F67–2, IIIC8, IIIE9 and IVD2), was done by deleting N- and C-terminal residues or by amino acid substitutions within these peptides. These studies revealed that the *M. leprae*-specific antibody IIIE9 and the crossreactive antibody IVD8 reacted to the same stretch of nine amino acids KLKLTGDEA of the *M. leprae* 65 kD protein (Figure 2.9). The difference in reactivity between these antibodies was explained by a different fine specificity: the essential residues for binding of IIIE9 appeared to be the K residue at position 1, the T residue at position 5 and the D, E, and A residues at positions 7–9, whereas for binding of IVD2 the K residue at position 1 and the E and A residues at positions 8 and 9 were essential. Thus, although in this case the defined sequence also contains a non-specific epitope with clear consequences for its use as a serodiagnostic reagent, in principle this strategy allows the dissection of amino acid sequences carrying specific B-cell epitopes that might be used as reagents for the serodiagnosis of mycobacterial infections.

```
426                      434
   K L K L T G D E A              IIIE9  :SPECIFIC

   K L K L T G D E A              IVD2   :CROSS-REACTIVE
```

Figure 2.9   Fine specificity of two *Mycobacterium leprae* B-cell epitopes on the 65 kD antigen (Brown *et al.*, 1986; Anderson *et al.*, 1988). Two monoclonal antibodies, one *M. leprae* specific (IIIE9) and one cross-reactive (IVD2), recognize the same peptide of the 65 kD protein. A small difference in the residues essential for binding (underlined) probably explains the different specificity of these antibodies.

4.4   T-cell epitopes

The mapping of T-cell epitopes on the 65 kD antigen was essentially done using the same tools as used for the mapping of B-cell epitopes, i.e. recombinant antigens and synthetic peptides (Lamb *et al.*, 1987; Oftung *et al.*, 1988; Thole *et al.*, 1988b; de Vries *et al.*, 1988; Lamb *et al.*, 1989; van Schooten *et al.*, 1989; Munk *et al.*, 1989; Anderson *et al.*, 1990). In some of these studies, this strategy was combined with prediction models for T-cell epitopes (Delisi and Berzofsky, 1985; Rothbard, 1986) So far such studies have led to the mapping of 21 T-cell epitopes (Figure 2.10). Both regions recognized by crossreactive or by mycobacterium-specific T cells were defined, and several of the conserved regions were shown to be identical on both the (myco)bacterial and the human 65 kD protein. One region, comprising residues 231–245, was recognized by both crossreactive and *M. tuberculosis* complex-specific T cells. The corresponding *M. leprae*

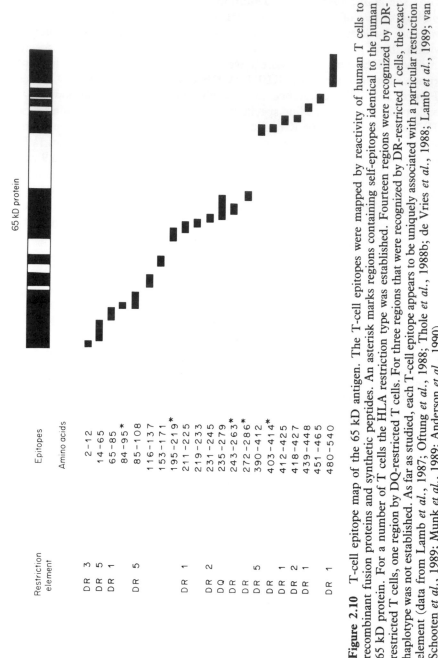

Figure 2.10 T-cell epitope map of the 65 kD antigen. The T-cell epitopes were mapped by reactivity of human T cells to recombinant fusion proteins and synthetic peptides. An asterisk marks regions containing self-epitopes identical to the human 65 kD protein. For a number of T cells the HLA restriction type was established. Fourteen regions were recognized by DR-restricted T cells, one region by DQ-restricted T cells. For three regions that were recognized by DR-restricted T cells, the exact haplotype was not established. As far as studied, each T-cell epitope appears to be uniquely associated with a particular restriction element (data from Lamb et al., 1987; Oftung et al., 1988; Thole et al., 1988b; de Vries et al., 1988; Lamb et al., 1989; van Schooten et al., 1989; Munk et al., 1989; Anderson et al., 1990).

231–245 sequence differs only in two residues, at positions 240 and 244, from the *M. tuberculosis*/*M. bovis* BCG sequence. Competition studies using the *M. leprae* peptide suggested that one or both of the residues 240 and 244 are involved in binding to MHC molecules on the antigen presenting cells, rather than in binding to the T-cell receptor of the *M. tuberculosis* complex-specific T cells. Thus, the specificity of this response is probably defined at the level of antigen presentation (Oftung *et al.*, 1988). The epitope-containing region recognized by one *M. leprae*-specific T-cell clone was studied in more detail and the epitope was defined to a sequence of 10 amino acids spanning the region 418–427 of the *M. leprae* protein (Anderson *et al.*, in press). T-cell responses to a large panel of peptides with amino acid substitutions at each position revealed that the amino acid residues 420–425 were essential for binding to the MHC molecule and/or recognition by the T-cell receptor.

For a number of the above mentioned MHC class-II-restricted CD4[+] T cells, responding in total to 15 distinct epitopes, the HLA phenotype restricting these T-cell responses was determined. From these 15 epitopes, 14 were shown to be recognized by DR-restricted T cells and one was recognized in the context of HLA-DQ (Figure 2.10). For three of the epitopes that were recognized by DR-restricted T cells, the exact haplotype was not yet established. The HLA restriction type of each of the T cells showed that each of the other 12 T-cell epitopes on the 65 kD protein is uniquely associated with a particular HLA phenotype: five are uniquely associated with HLA-DR1, two with HLA-DR2, one with HLA-DR3, three with HLA-DR5, and one with HLA-DQ. In mice, this unique association between T-cell epitopes and MHC phenotype was not found, but in these animals, too, the T-cell response to particular epitopes on the 65 kD antigen appears to be highly influenced by the MHC haplotype (Brett *et al.*, 1989). Thus, the MHC class II molecules seem to play an important role in the presentation of particular epitopes of the 65 kD protein to the immune system, and in general this observation has to be taken into account in the design of subunit vaccines.

5   THE ROLE OF THE 65 kD ANTIGEN IN AUTO-IMMUNE ARTHRITIS

To study the pathogenesis of arthritis, several animal models are in use in which the disease is evoked by immunization with bacterial antigens or with synthetic compounds. In Lewis rats so-called adjuvant arthritis (AA) develops about 14 days after intradermal injection of heat-killed *M. tuberculosis* suspended in mineral oil, a preparation known as Complete Freund's Adjuvant (Pearson, 1964). Studies initiated by Irun Cohen at the Weizman Institute of Science have shown that a T-cell line, A2, obtained from a Lewis

rat with adjuvant arthritis and cultured *in vitro* with *M. tuberculosis* as
the selecting antigen, could transfer the disease to irradiated, naive rats
(Holoshitz *et al.*, 1983). In non-irradiated animals inoculation of line A2
appeared to vaccinate against the induction of AA. Two clones derived from
this line were found to possess only one of these opposed functionalities:
treatment with clone A2b produced the disease in irradiated rats, while clone
A2c protected against subsequent attempts to induce AA (Holoshitz *et al.*,
1984).

Analysis with several recombinant mycobacterial antigens revealed that

Table 2.2 Mapping of the Adjuvant Arthritis Associated T-Cell Epitope in the
Mycobacterial 65 kD Protein. All 28 overlapping peptides in the 170–205 region of
the 65 kD protein were prepared by multiple peptide synthesis and their stimulatory
activity (SI) with clones A2b and A2c was tested. A high proliferative response was
found only with peptides containing the 180–186 sequence (Van der Zee *et al.*, 1989).

170 180 190 200	*T-cell response*	*(SI)*
EGVITVEESNTFGLQLELTEGMRFDKGYISGYFVTD	A2b	A2c
EGVITVEES	<2	<2
GVITVEESN	<2	<2
VITVEESNT	<2	<2
ITVEESNTF	<2	<2
TVEESNTFG	<2	<2
VEESNTFGL	<2	<2
EESNTFGLQ	<2	<2
ESNTFGLQL	63	14
SNTFGLQLE	773	440
NTFGLQLEL	812	413
TFGLQLELT	616	395
FGLQLELTE	<2	<2
GLQLELTEG	<2	<2
LQLELTEGM	<2	<2
QLELTEGMR	<2	<2
LELTEGMRF	<2	<2
ELTEGMRFD	<2	<2
LTEGMRFDK	<2	<2
TEGMRFDKG	<2	<2
EGMRFDKGY	<2	<2
GMRFDKGYI	<2	<2
MRFDKGYIS	<2	<2
RFDKGYISG	<2	<2
FDKGYISGY	<2	<2
DKGYISGYF	<2	<2
KGYISGYFV	<2	<2
GYISGYFVT	<2	<2
YISGYFVTD	<2	<2
M. bovis 65 kD protein (5 µg/ml)	589	557

both clones recognized the *M. tuberculosis/M. bovis* BCG 65 kD protein (Van Eden *et al.*, 1988). Further analysis with fusion proteins and synthetic peptides showed that both clones were fully stimulated by one and the same peptide comprising residues 180–186 of the 65 kD protein (Van Eden *et al.*, 1988; Van der Zee et al., 1989; Table 2.2). In addition to their specificity for this mycobacterial peptide, both clones give a proliferative response to proteoglycan preparations and to human synovial fluid. Thus, antigenic mimicry between the immunizing antigen and an endogenous antigen as recognized by a virulently arthritogenic T-cell clone explains the auto-immune nature of adjuvant arthritis (Van Eden *et al.*, 1985, 1987).

It is tempting to assume that the crossreactive antigen is the rat homologue of the 65 kD protein. The amino acid sequence of the 65 kD proteins are highly conserved, and the protein is present in inflamed human joints (McLean *et al.*, 1988). However, the 180–186 region appears to be one of the more variable parts of the protein. Out of the seven residues only three are identical in the human and in the mycobacterial protein (Jindal *et al.*, 1989). Preliminary results on the relevant sequence in the Lewis rat 65 kD protein do not reveal a higher degree of identity with the mycobacterial 180–186 sequence (Van der Zee, unpublished data). It therefore seems unlikely that the Lewis 65 kD protein is the crossreactive target that is recognized by clones A2b and A2c in rat tissue, unless other yet unknown forms of this protein are expressed by homologous genes, possibly in a tissue-specific way. As stated above, clones A2b and A2c, with their indistinguishable receptor specificity were found to have opposing effects *in vivo*. This suggests that recognition of the mycobacterial 180–186 sequence by the immune system does not necessarily lead to adjuvant arthritis in Lewis rats. In fact, it was shown that pre-immunization of rats with purified 65 kD protein induced resistance to subsequent attempts to induce the disease (Van Eden *et al.*, 1988). Probably, without the presence of mycobacterial adjuvant components, a more A2c-like response is induced.

Interestingly, an equally important role of the 65 kD protein was recently demonstrated in other models of experimental arthritis. A chronic form of polyarthritis can be induced in Lewis rats by immunization with a preparation of streptococcal cell walls [SCW] (Chromartie *et al.*, 1977; Van den Broek *et al.*, 1988). In this model too, a mimicry at the level of T cells between the disease-inducing compound, SCW, and cartilage was found and, as in the AA model, the disease can be transferred by T cells to irradiated naive animals (Quinn Dejoy *et al.*, 1989). Recently, it was shown that also in the SCW model, pre-administration of the mycobacterial 65 kD protein at no more than five days prior to immunization with SCW, completely blocked the development of arthritis (Figure 2.11). This suppressive effect seemed to be antigen-specific, as the response to irrelevant antigens was not influenced

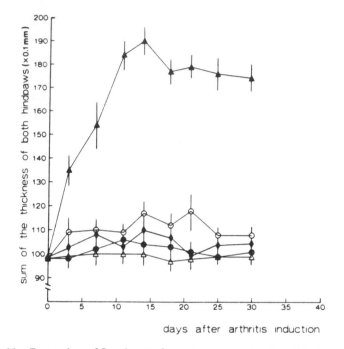

days after arthritis induction

Figure 2.11   Protection of Lewis rats from streptococcal cell wall induced arthritis by pretreatment with 50 μg of 65 kD protein of *M. bovis* BCG in incomplete Freund adjuvant. Rats were pretreated intraperitoneally at 35 (◆), 25 (◇), 15 (●), or 5 days (○) before arthritis induction. The control group received incomplete Freund adjuvant alone at day -25 (▲). Arthritis was scored by measuring the thickness of both hindpaws (Van den Broek *et al.*, 1989).

and crossreactivity between the 65 kD protein and SCW antigens was demonstrated by antibody and T-cell reactivity (Van den Broek *et al.*, 1989). On the other hand, Quinn Dejoy *et al.* (1989) did not observe proliferation of an arthritogenic T-cell line in response to the recombinant *M. tuberculosis* 65 kD antigen in the SCW model.

   Likewise, in a third model in which the disease-inducing agent is of non-bacterial origin, the 65 kD antigen was also shown to protect against arthritis. The induction of arthritis in mice by pristane, a synthetic adjuvant, was effectively suppressed by pre-injection with the mycobacterial protein (Thompson *et al.*, 1990). It is clear from this observation that the protective effect of the protein is not necessarily a consequence of the presence of crossreactive structures in both the protective 65 kD protein and the disease-inducing material, but more likely a consequence of a direct mimicry with endogenous target structures. Obviously, the 65 kD protein may be a common denominator in different experimental models of arthritis.

As regards humans, a few studies suggest involvement of the protein in auto-immune arthritis as well. Its behaviour as a major immunogen in human beings was described already in the previous section. Elevated antibody levels against the 65 kD protein of mycobacteria were found in sera obtained from rheumatoid arthritis patients (Bahr *et al.*, 1988). In addition, increased T-cell responses to the protein were observed in patients in the early onset of the disease, and were more prominent in synovial fluid lymphocytes than in peripheral blood lymphocytes (Res *et al.*, 1988). Analogous high responses to the 65 kD protein of mycobacteria but not to the homologue of *E. coli*, were found in synovial lymphocytes of patients with reactive forms of arthritis (Gaston *et al.*, 1989).

Summarizing, a rather special relationship between the 65 kD protein and different forms of arthritis both in animal and in humans seems to be becoming evident.

6 CONCLUDING REMARKS

In recent years recombinant DNA technology has allowed considerable progress in the analysis of mycobacterial antigens. The 65 kD antigen was the most thoroughly investigated and several features became apparent with implications reaching beyond the field of mycobacteriology. The discovery that this antigen belongs to a family of heat-shock proteins with counterparts probably in each cell implicated a highly conserved functional and immunological role for this protein. Detailed analysis of both prokaryotic and eukaryotic members of this family showed that they function in the assembly and disassembly of protein complexes, and that they are important for proper transport of certain proteins through cell membranes. Their increased production in response to heat shock and other stress conditions probably reflects a mechanism of maintaining functional conformation of proteins essential for the cell during adverse conditions.

The conservation of this antigen throughout nature has provided the immune system with a broad spectrum of epitopes present on many pathogens other than mycobacteria and even self-epitopes present on autologous cells. The frequent recognition of these conserved epitopes by both patients and healthy individuals indicated an important immunological role for the 65 kD antigen not only in diseased persons, but also in healthy subjects. The immune response to the 65 kD antigen may represent an antigen-specific first line of defence in the elimination of infectious pathogens as well as in the elimination of stressed, autologous cells. Defects in the regulation of this response to either pathogenic or autologous cells may possibly lead to autoimmune disease such as rheumatoid arthritis.

ACKNOWLEDGEMENTS

We would like to thank W. van Schooten, J. van Embden, T. Ottenhoff, R. Hartskeerl, P. Klatser, C. Verstijnen and A. Kolk for their helpful discussion and K. Boven and M. Tensen for help in preparing the manuscript.

REFERENCES

Ab, K. B., Kiessling, R., Van Embden, J. D. A., Thole, J. E. R., Kumararatne, D. S., Pisa, P., Wondimu, A. and Ottenhoff, T. H. M. (1990). *Eur. J. Immunol.* **20**, 369–377.

Andersen, A. S., Worsaae, A. and Chaparas, S. D. (1988). *Infect. Immun.* **56**, 1344–51.

Anderson, D. C., Barry, M. E. and Buchanan, T. M. (1988). *J. Immun.* **141**, 607–13.

Anderson, D. C., Van Schooten, W. C. A., Janson, A., Barry, M. E. and De Vries, R. R. P. (1990). *J. Immunol.* **144**, 2459–2464.

Athwal, R. S., Deo, S. S. and Imaeda T. (1984). *Int. J. Syst. Bacteriol.* **34**, 371–5.

Bahr, G. M., Rook, G. A. W., Al-Saffar, M., van Embden, J., Stanford, J. L. and Behbehani, K. (1988). *Clin. Exp. Immunol.* **74**, 211–15.

Baird, P. N., Hall, L. M. C. and Coates, A. R. M. (1989). *J. Gen. Microbiol.* **135**, 931–9.

Barksdale, L. and Kim, K. (1977). *Bacteriol. Rev.* **4**, 217–372.

Bochkareva, E. S., Lissin, N. M. and Girshovich, A. S. (1988). *Nature* **336**, 254–7.

Boom, W. H., Husson, R. N., Young, R. A., David, J. R. and Piessens, W. F. (1987). *Infect. Immun.* **55**, 2223–9.

Brett, S. J., Lamb, J. R., Cox, J. H., Rothbard, J. B., Mehlert, A. and Ivanyi, J. (1989). *Eur. J. Immun.* **19**, 1303–10.

Brown, W. B., Larrabee, W. A. and Kim, P. S. (1986). *Lepr. Rev.* **57**, (Suppl. 2) 157–62.

Buchanan, T. M., Nomaguchi, H., Anderson, D. C., Young, R. A., Gillis, T. P., Britton, W. J., Ivanyi, J., Kolk, A. H. J., Closs, O., Bloom, B. R. and Mehra V. (1987). *Infect. Immun.* **55**, 1000–3.

Chanda, P. K., Ono, M., Kuwano, M. and Kung, H.-F. (1985). *J. Bacteriol.* **161**, 446–9.

Chandrasekhar, C. N., Tilly, K., Woolford, C., Hendrix, R. and Georgopoulos, C. (1986). *J. Biol. Chem.* **261**, 12414–19.

Cheng, M. Y., Hartl, F. U., Martin, J., Pollock, R. A., Kalousek, F., Neupert, W., Hallberg, E. M., Hallberg, R. L. and Horwich, A. L. (1989). *Nature* **337**, 620–5.

Chromartie, W. J., Craddock, J. G., Schwab, J. H., Anderle, S. K. and Yang, C. (1977). *J. Exp. Med.* **146**, 1585–602.

Clark-Curtiss, J. E., Jacobs, W. R., Docherty, M. A., Ritchie, L. R. and Curtiss III, R. (1985). *J. Bacteriol.* **161**, 1093–102.

Closs, O., Harboe, M., Axelsen, N. H., Bunch-Christensen, K. and Magnusson, M. (1980). *Scand. J. Immunol.* **12**, 249–63.

Coates, A. R. M., Allen, B. W., Hewitt, J., Ivanyi, J. and Mitchison, D. A. (1981). *Lancet* ii, 167–9.

Daniel, T. M. and Janicki, B. W. (1978). *Microbiol. Rev.* **42**, 84–113.

De Bruyn, J., Bosmans, R., Turneer, M., Weckx, M. Nyabenda, J., Van Vooren, J.-P., Falmagne, P., Wiker, H. G. and Harboe, M. (1987). *Infect. Immun.* **55**, 245–52.

De Bruyn, J., Bosmans, R., Nyabenda, J. and Van Vooren, J. P. (1989). *J. Gen. Microbiol.* **135**, 79–84.

Delisi, G. and Berzofsky, J. A. (1985). *Proc. Natl. Acad. Sci. USA* **82**, 7048–52.

De Vries, R. R. P., Ottenhoff, T. H. M. and Van Schooten, W. C. A. (1988). *Springer. Semin. Immunopathol.* **10**, 305–18.

Ellis, J. (1987). *Nature* **328**, 378–9.

Emmrich, B., Thole, J., Van Embden, J. and Kaufmann, S. H. E. (1986). *J. Exp. Med.* **163**, 1024–9.

Engers, H. D., et al. (1985). *Infect. Immun.* **48**, 603–5.

Engers, H. D., *et al.* (1986). *Infect. Immun.* **51**, 718–20.

Fayet, O., Louarn, J.-M. and Georgopoulos, C. (1986). *Mol. Gen. Genet.* **202**, 435–45.

Gaston, J. H. S., Life, P. F., Bailey, L. C., and Bacon, P. A. (1989). *J. Immunol.* **143**, 2494–500.

Georgopoulos, C. P., Hendrix, R. W., Casjens, S. R. and Kaiser, A. D. (1973). *J. Mol. Biol.* **76**, 45–60.

Gillis, T. P., Miller, R. A., Young, D. B., Khanolkar, S. R. and Buchanan, T. M. (1985). *Infect. Immun.* **49**, 371–7.

Goloubinoff, P., Gatenby, A. and Lorimer, G. H. (1989). *Nature* **337**, 44–7.

Goren, M. B. (1982) *Am. Rev. Respir. Dis.* **125**, 50–69.

Hance, A. J., Grandchamp, B., Lévy-Frebault, V., Lecossier, D., Rauzier, J., Bocart, D. and Gicquel, B. (1989). *Mol. Microbiol.* **3**, 843–9.

Hansen, K., Bangsborg, J. M., Fjordvang, H., Strandberg Pedersen, N. and Hinersson, P. (1988). *Infect. Immun.* **56**, 2047–53.

Harboe, M. and Wiker, H. G. (1986). *Lepr. Rev.* **57** (Suppl. 2), 33–7.

Haregewoin, A., Soman, G., Hom, R. C. and Finberg, R. W. (1989). *Nature* **340**, 309–12.

Hemmingsen, S. M., Woolford, C., Van der Vies, S. M., Tilly, K., Dennis, D. T., Georgopoulos, C. P., Hendrix, R. W. and Ellis, R. J. (1988). *Nature* **333**, 330–4.

Hendrix, R. W. (1979). *J. Mol. Biol.* **129**, 375–92.

Higgins, D. G. and Sharp, P. M. (1988). *Gene* **73**, 237–44.

Higgins, D. G. and Sharp, P. M. (1989). *Cabios* **5**, 151–3.

Hindersson, P., Knudsen, J. D. and Axelsen, N. H. (1987). *J. Gen. Microbiol.* **133**, 587–96.

Hohn, T., Hohn, B., Engel, A., Wurtz, M. and Smith, P. R. (1979). *J. Mol. Biol.* **129**, 359–73.

Hoiby, N. (1975a). *Scand. J. Immunol.* **4** (Suppl. 2), 187–96.

Hoiby, N. (1975b). *Scand. J. Immunol.* **4** (Suppl. 2), 197–202.

Holoshitz, J., Naparstek, Y., Ben-Nun, A. and Cohen, I. R. (1983). *Science* **219**, 56–8.

Holoshitz, J., Matitiau, A. and Cohen, I. R. (1984). *J. Clin. Invest.* **73**, 211–15.

Holoshitz, J., Koning, F., Coligan, J. E., De Bruyn, J. and Strober, S. (1989). *Nature* **339**, 226–9.

Hunter, S. W., McNeil, M., Modlin, R. L., Mehra, V., Bloom, B. R. and Brennan, P. J. (1989). *J. Immunol.* **142**, 2864–72.

Husson, R. N. and Young R. A. (1987). *Proc. Natl. Acad. Sci. USA* **84**, 1679–83.

Ivanyi, J., Morris, J. A., Keen, M. (1985). In *Monoclonal Antibodies Against Bacteria* (eds A. J. L. Macario and E. C. Macario), pp. 59–90. Academic Press, NY.

Jacobs, W. R., Docherty, R., Curtiss III, R. and Clark-Curtiss, J. E. (1986). *Proc. Natl. Acad. Sci. USA* **83**, 1926–30.

Jenkins, J. A., March, J. B., Oliver, I. R. and Masters, M. (1986). *Mol. Gen. Genet.* **202**, 446–54.

Jindal, S., Dudani, A. K., Singh, B., Harley, C. B. and Gupta, R. S. (1989). *Mol. Cell. Biol.* **9**, 2279–83.

Kaijser, B. (1975). *Inter. Archs. Allergy Appl. Immunol.* **48**, 72–81.

Kaufmann, S. H. E., Väth, U., Thole, J. E. R., Van Embden, J. D. A. and Emmrich, F. (1987). *Eur. J. Immun.* **17**, 351–7.

Kochan, J. and Murialdo, H. (1983). *Virology* **131**, 100–15.

Lamb, F. I., Kingston, A. E., Estrada-G., I. and Colston, M. J. (1988). *Infect. Immun.* **56**, 1237–41.

Lamb, J. R., Ivanyi, J., Rees, A., Young, R. A. and Young, D. B. (1986). *Lepr. Rev.* 57 (Supp. 2), 131–137.

Lamb, J. R., Ivanyi, J., Rees, A. D. M., Rothbard, J. B., Howland, K., Young, R. A. and Young, D. B. (1987). *EMBO Journal* 6, 1245–9.

Lamb, J. R., Bal, V., Mendez-Samperio, P., Mehlert, A., So, A., Rothbard, J., Jindal, S., Young, R. A. and Young, D. B. (1989). *Int. Immunol.* 1, 191–6.

Lindquist, S. (1986). *Ann. Rev. Biochem.* 55, 1151–91.

Lindquist, S. and Craig, E. A. (1988). *Ann. Rev. Genet.* 22, 631–7.

Lu, M. C., Lien, M. H., Becker, R. E., Heine, H. C., Buggs, A. M., Lipovsek, D., Gupta, R., Robbins, P. W., Grosskinsky, C. M., Hubbard, S. C. and Young, R. A. (1987). *Infect. Immun.* 55, 2378–82.

McLean, I. L., Winrow, V. R., Mapp, P. I., Cherrie, A. H., Archer, J. R. and Blake, D. R. (1988). *Lancet* ii, 856–7.

McMullin, T. W. and Hallberg, R. L. (1988). *Mol. Cell. Biol.* 8, 371–80.

Mehra, V., Sweetser, D. and Young, R. A. (1986). *Proc. Natl. Acad. Sci. USA* 83, 7013–17.

Miziorko, H. M. and Lorimer, G. H. A. (1983), *Rev. Biochem.* 52, 507–35.

Morrison, R. P., Belland, R. J., Lyng, K. and Caldwell, H. D. (1989). *J. Exp. Med.* 163, 1024–9.

Munk, M. E., Schoel, B. and Kaufmann, S. H. E. (1988). *Eur. J. Immunol.* 18, 1835–8.

Munk, M. E., Schoel, B., Modrow, S., Karr, R. W., Young, R. A. and Kaufmann, S. H. E. (1989). *J. Immunol.* 143, 2844–9.

Mustafa, A. S., Gill, H. K., Nerland, A., Britton, W. J., Mehra, V., Blooms, B. R., Young, R. A. and Godal, T. (1986). *Nature* 319, 63–8.

Neidhardt, F. C., Phillips, T. A., VanBogelen, R. A., Smith, M. W., Georgalis, Y. and Subramanian, A. R. (1981). *J. Bacteriol.* 145, 513–20.

O'Brien, R. L., Pat Happ, M., Dallas, A., Palmer, E., Kubo, R. and Born, W. K. (1989). *Cell* 57, 667–74.

Oftung, F., Mustafa, A. S., Husson, R., Young, R. A. and Godal, T. (1987). *J. Immunol.* 138, 927–31.

Oftung, F., Mustafa, A. S., Shinnick, T. M., Houghten, R. A., Kvalheim, G., Degre, M., Lundin, K. E. A. and Godal, T. (1988). *J. Immunol.* 141, 2749–54.

Ottenhoff, T. H. M., Kale Ab, B., Van Embden, J. D. A., Thole, J. E. R. and Kiessling, R. (1988). *J. Exp. Med.* 168, 1947–52.

Pau, C. P., Plikaytis, B. B., Carlone, G. M. and Warner, I. M. (1988). *J. Clin. Microbiol.* 26, 67–71.

Pearson, C. M. (1964). *Arthritis Rheum.* 7, 80–6.

Puskin, A. C., Tsuprin, V. L., Solovjeva, N. A., Shubin, V. V., Evstigneeva, Z. G. and Kretsovisch W. L. (1982). *Biochim. Biophys. Acta* 704, 379–84.

Quinn Dejoy, S., Ferguson, K. M., Sapp, T. M., Zabriskie, J. B., Oronsky, A. L. and Kerwar, S. S. (1989). *J. Exp. Med.* 170, 369–82.

Raulet, D. H. (1989). *Nature* 339, 342–3.

Reading, D. S., Hallberg, R. L. and Myers, A. M. (1989). *Nature* 337, 665–9.

Res, P. C. M., Schaar, C. G., Breedveld, F. C., van Eden, W., van Embden, J. D. A., Cohen, I. R., de Vries, R. R. P. (1988). *Lancet* ii, 478–80.

Rothbard, J. B. (1986). *Ann. Inst. Pasteur.* 137E, 518–26.

Sand Petersen, C., Strandberg Pedersen, N. and Axelsen N. H. (1982). *Scand. J. Immunol.* 15, 459–65.

Shinnick, T. M. (1987). *J. Bacteriol.* 169, 1080–8.

Shinnick, T. M., Sweetser, D., Thole, J., Van Embden, J. and Young R. A. (1987). *Infect. Immun.* 55, 1932–5.

Shinnick, T. M., Vodkin, M. H. and Williams J. C. (1988). *Infect. Immun.* **56**, 446–51.

Sompolinsky, D., Hertz, J. B., Hoiby, N., Jensen, K., Mansa, B., Pedersen, V. B. and Samra Z. (1980). *Acta Pathol. Microbiol. Scand. Sect. B.* **88**, 253–60.

Sternberg, N. (1973). *J. Mol. Biol.* **76**, 25–44.

Thole, J. E. R., Dauwerse, H. G., Das, P. K., Groothuis, D. G., Schouls, L. M. and Van Embden, J. D. A. (1985). *Infect. Immun.* **50**, 800–6.

Thole, J. E. R., Keulen, W. C., De Bruyn, J., Kolk, A. H. J., Groothuis, D. G., Berwald, L. G., Tiesjema, R. H. and Van Embden, J. D. A. (1987). *Infect. Immun.* **55**, 1466–75.

Thole, J. E. R., Hindersson, P., De Bruyn, J., Cremers, F., Van der Zee, J., De Cock, H., Tommassen, J., Van Eden W. and Van Embden, J. D. A. (1988a). *Microbial Pathogenesis* **4**, 71–83.

Thole, J. E. R., Van Schooten, W. C. A., Keulen, W. J., Hermans, P. W. M., Janson, A. A. M., De Vries, R. R. P., Kolk, A. H. J. and Van Embden, J. D. A. (1988b). *Infect. Immun.* **56**, 1633–40.

Thompson, S. J., Bedwell, A., Hooper, D. C., Burtless, S. S., Rook, G. A. W. and Elson, C. J. (1990). In *Stress Proteins in Inflammation* (eds C. Rice-Evans, V. Winrow, D. Blake and R. Burdon). In press. Richelieu Press.

Van den Broek, M. F., van de Berg, W. B., Arntz, O. J. and van de Putte, L. B. A. (1988). *Clin. Exp. Immunol.* **72**, 9–14.

Van den Broek, M. F., Hogervorst, E. J. M., Van Bruggen, M. C. J., Van Eden, W., Van der Zee, R. and Van den Berg, W. B. (1989). *J. Exp. Med.* **170**, 449–66.

Van Eden, W., Holoshitz, J., Nevo, Z., Frenkel, A., Klajman, A. and Cohen, I. R. (1985). *Proc. Natl. Acad. Sci. USA* **82**, 5117–19.

Van Eden, W., Holoshitz, J. and Cohen, I. R. (1987). *Concepts Immunopathol.* **4**, 144–70.

Van Eden, W., Thole, J. E. R., Van der Zee, R., Noordzij, A., Van Embden, J. D. A., Hensen, E. J. and Cohen, I. R. (1988). *Nature* **331**, 171–3.

Van Schooten, W. C. A., Ottenhoff, T. H. M., Klatser, P. R., Thole, J. E. R., De Vries, R. R. P. and Kolk, A. H. J. (1988). *Eur. J. Immunol.* **18**, 849–54.

Van Schooten, W. C. A., Elferink, D. G., Van Embden, J., Anderson, D. C. and De Vries, R. R. P. (1989). *Eur J. Immunol.* **19**, 2075–9.

Van der Zee, R., Van Eden, W., Meloen, R. H., Noordzij, A. and Van Embden, J. D. A. (1989). *Eur. J. Immunol.* **19**, 43–7.

Vodkin, M. H. and Williams, J. C. (1988). *J. Bacteriol.* **170**, 1227–34.

Watson, J. D. (1989). *Immunology Today* **10**, 218–21.

Wilske, B., Preac-Mursic, V., Schierz, G. and Busch, V. (1986). *Zentralbl. Bakteriol. Hyg. A.* **263**, 92–102.

Young, D. B. (1988). *Brit. Med. Bull.* **44**, 562–583.

Young, D. B., Ivanyi, J., Cox, J. H. and Lamb, J. R. (1987a). *Immunology Today* **8**, 215–18.

Young, D. B., Kent, L. and Young, R. A. (1987b). *Infect. Immun.* **55**, 1412–25.

Young, D. B., Lathigra, R., Hendrix, R., Sweetser, D. and Young, R. A. (1988). *Proc. Natl. Acad. Sci. USA* **85**, 4267–70.

Young, R. A. (1990). *Annu. Rev. Immunol.* (in press).

Young, R. A. and Davis, R. W. (1983). *Proc. Natl. Acad. Sci. USA* **80**, 1194–8.

Young, R. A. and Elliot, T. (1989). *Cell* **59**, 5–8.

Young, R. A., Blooms, B. R., Grosskinsky, C. M., Ivanyi, J., Thomas, D. and Davis, R. W. (1985a). *Proc. Natl. Acad. Sci. USA* **42**, 2583–2587.

Young, R. A., Mehra, V., Sweetser, D., Buchanan, T., Clark-Curtiss, J., Davis, R. W. and Bloom, B. R. (1985b). *Nature* **316**, 450–2.

Zweig, M. and Cummings, D. (1973). *J. Mol. Biol.* **80**, 505–18.

3 Protective immunity against mycobacterial infections: investigating cloned antigens

M. J. Colston

Laboratory for Leprosy and Mycobacterial Research, National Institute for Medical Research, The Ridgeway, Mill Hill, London NW7 1AA, UK

1 INTRODUCTION: THE NATURE OF PROTECTIVE IMMUNITY IN INTRACELLULAR INFECTIONS

Mycobacterial infections are characterized by their intracellular nature. The pathogenic mycobacteria have evolved mechanisms for surviving and prospering within cells of the reticuloendothelial system, and, in the case of *Mycobacterium leprae*, Schwann cells of peripheral nerves. In general, intracellular infections are conrolled by cell-mediated mechanisms, rather than by antibody responses, and this appears to hold true for the mycobacteria. Thus *M. leprae* infections in congenitally athymic mice, which lack T cells, are greatly enhanced compared to those in normal mice (Colston and Hilson, 1976), and *M. tuberculosis* infections in normal mice can be controlled by the transfer of immune T cells (Orme and Collins, 1986).

The exact nature of the T cells involved in protective immunity to mycobacteria is, as yet, uncertain. For functional purposes, T lymphocytes have been divided into two major subsets which correspond to the presence of surface markers on the cell and to the nature of the major histocompatibility complex (MHC) molecule which is recognized in association with the foreign antigen. In general, helper T cells (T_H) possess a surface marker termed CD4 (L3T4 in the mouse) and recognize foreign antigen in association with MHC Class II molecules. Cytotoxic T cells (Tc) possess a surface marker termed

MOLECULAR BIOLOGY OF THE MYCOBACTERIA
ISBN 0–12–483378–0

CD8 (Lyt 2 in the mouse) and recognize foreign antigen in association with Class I molecules (for review see Fitch, 1986). However, these associations between function and phenotype are not absolute; for example T cells of helper phenotype (CD4$^+$, Class II restricted) have been shown to exhibit antigen specific cytotoxic activity (Ottenhoff et al., 1988). The role of these different T-cell subsets and the mechanisms by which they mediate immunity against intracellular infectious agents are unresolved issues of much debate and controversy (see D. Young et al., this volume).

The conventional view of development of immunity to intracellular bacteria is derived mainly from the studies of Mackaness and colleagues working with Listeria monocytogenes. Sensitized T cells, on recognition of antigen in association with MHC molecules on the surface of infected macrophages, release cytokines which stimulate the macrophages to express enhanced antimicrobicidal activity. In addition, the sensitized T cells themselves perform a number of functional roles other than the activation of macrophages, including the release of cytokines that promote the migration of monocytes into the site of bacterial invasion thus contributing to the formation of a granuloma, the mediation of cutaneous delayed-type hypersensitivity, and the retention of a long-lived state of immunological memory.

Several recent observations have served to question this dogma. These include the demonstration of T cells with cytotoxic activity against mycobacteria-primed T cells, and the difficulty of demonstrating in vitro activation of macrophages to kill mycobacteria. Even more recently, information on the T-cell response following immunization with mycobacteria has created a new possibility; the vast majority of T cells express T-cell receptors made up of α and β chains. However, in mice which are immunized with M. tuberculosis there is a large increase in T cells expressing a different T-cell receptor, made up of γ and δ chains (Janis et al., 1989). T cells expressing γ δ receptors do not express either CD4 or CD8 molecules, and their function is unknown. However, in addition to the work in immunized mice, γ δ bearing T cells have been found to proliferate in response to mycobacterial antigens (Holoshitz et al., 1989; O'Brien et al., 1989; Modlin et al., 1989), and to be increased by up to eightfold in lesions from leprosy patients undergoing reversal reactions, and in lepromin skin tests (Modlin et al., 1989). At present the role of these cells in immunity to the mycobacteria is purely conjectural, but a role in resistance to disease or in the immunopathology of mycobacteral infections cannot be ruled out.

2   ANIMAL MODELS FOR INVESTIGATING PROTECTIVE IMMUNITY

In leprosy, the animal model for studying protective immunity is simplified by a lack of choice. The only system which has been used to any degree is that which involves testing the effect of immunization on the growth of *M. leprae* in the footpads of mice (Shepard *et al.*, 1980; Singh *et al.*, 1989). Small numbers of *M. leprae* (usually between 5000 and 10 000) are inoculated into the hind footpads of mice; after approximately six months the numbers of bacilli within the footpads have increased by approximately 100-fold. If mice are immunized, prior to inoculation in the footpad, with killed *M. leprae* given intradermally, the footpad infection is inhibited (Figure 3.1).

Several animal models can be used for investigating protection against *M. tuberculosis*. Much work is carried out in mice where infection can be by intravenous, intraperitoneal or intranasal routes. Guinea pigs are highly susceptible to infection with *M. tuberculosis;* subcutaneous or intramuscular injection of even small numbers of bacilli result in systemic spread of infection and, ultimately, death to the animal.

3 THE ANTIGENS RECOGNIZED BY MURINE T CELLS

It is generally true that T cells recognize peptide fragments produced from processed proteins. Although examples of T cells recognizing non-protein molecules exist in the literature, (Mehra *et al.*, 1984), this appears to be a rare phenomenon. Since T cells recognize proteins and T cells are the functionally important cells in priming protective immunity, it seems reasonable to believe that proteins play a central role in mediating protective immunity. However, a contributory role by other molecules cannot be ruled out and in a structure as complex as a mycobacterial cell it is likely that many constituents are involved; many mycobacterial components are known, for example, to have powerful adjuvant properties which will contribute to the overall immune response.

The available evidence suggests that when mice mount an immune response against mycobacteria, T-cell responses against many different mycobacterial proteins can be detected. Kingston *et al.*, (1986), immunizing mice intradermally with killed *M. leprae*, a procedure known to generate protective immunity, produced T-cell clones with many different recognition specificities. Similar results have been obtained with human T-cell clones (Ottenhoff *et al.*, 1986). On the other hand certain proteins are probably recognized more frequently than others. It has been estimated, for example, that in mice immunized with *M. tuberculosis*, approximately 20% of mycobacteria-reactive T cells recognize the 65 kD protein (Kaufman *et al.*, 1987). Similarly, in *M. leprae*-immunized mice the 65 kD protein is a major site for T-cell recognition (Lamb *et al.*, 1988). However, the available evidence suggests that, even if one analyses T-cell responses to a single protein, a number of different T cell epitopes are recognized (see Thole *et al.*,

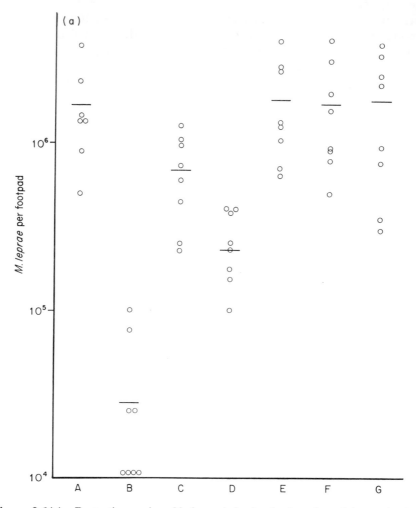

Figure 3.1(a) Protection against *M. leprae* infection by intradermal immunization. Immunization was with A, non-immunized controls; B, killed *M. leprae*; C, purified 65 kD protein; D, viable *E. coli* expressing 65 kD protein; E, viable *E. coli* control; F, viable *Streptomyces lividans* expressing 65 kD protein; G, viable *S. lividans* control.

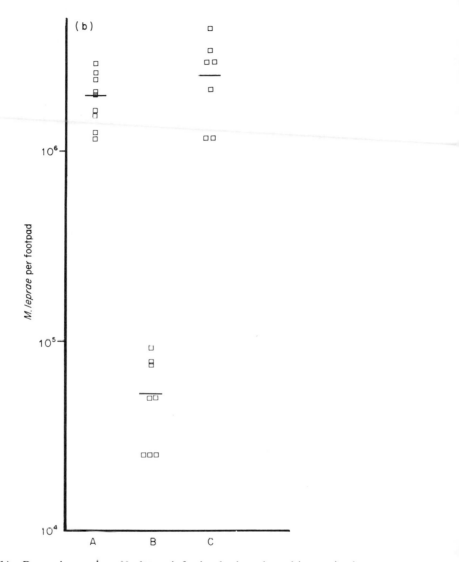

Figure 3.1(b)   Protection against *M. leprae* infection by intradermal immunization. Immunization was with A, non-immunized control; B, killed *M. leprae*; C, viable *E. coli* expressing 65 kD protein.

this volume), and hence unravelling which epitopes are important in protective immunity will be a considerable task.

When studying the molecular structure of mycobacterial antigens, attention has been focused on those antigens which exhibit some degree of species specificity. As far as protection against *M. leprae* infection is concerned, immunization with *M. leprae* is more successful at protecting mice against infection with *M. leprae* than immunization with other mycobacteria, although BCG (Shepard *et al.*, 1980; Singh *et al.*, 1989) and another cultivable mycobacterium, *M. habana* (Singh *et al.*, 1989 and see below) also give protection in the mouse footpad system. With *M. tuberculosis* infection, significant protection can be conferred with crossreactive mycobacteria (Orme and Collins, 1986), hence questioning the requirement for specificity.

4   INVESTIGATIONS WITH CLONED ANTIGENS

With the production of cloned mycobacterial proteins, the obvious next step was to immunize mice with recombinant proteins and investigate the effect on the growth of mycobacteria. The first protein available in large amounts, the *M. bovis* 65 kD protein (Shinnick, 1987; Thole *et al.*, 1987) and the *M. leprae* 65 kD protein have been investigated in this way by a number of groups.

In our laboratory, we have immunized mice with *M. leprae* r65 kD protein given in a variety of forms. Figure 1(a) shows an experiment in which mice were immunized with purified r65 kD protein, with recombinant *E. coli* expressing the protein at a high level, and with *Streptomyces lividans* also expressing the protein. The positive controls (mice immunized with killed *M. leprae*) show little growth of *M. leprae* on subsequent challenge with viable organisms in the mouse footpad when compared to non-immunized controls. None of the groups immunized with r65 kD showed protection equivalent to that seen with *M. leprae*. Mice immunized with *E.coli* expressing the protein showed some reduction in counts compared to unvaccinated controls, but when this was repeated (Figure 1(b)), no protection could be detected.

Subsequent experiments immunizing mice with 65 kD protein suggest that it may actually have an inhibitory effect on the immunization of mice with killed *M. leprae*. When r65 kD is combined with *M. leprae* there is a partial, but significant, abrogation of the protective effect. This effect has now been observed in several experiments, and the mechanisms involved are currently under investigation. It appears to be similar to the tolerance induction by this protein in rats subsequently challenged to develop adjuvant arthritis (Billingham *et al.*, 1990) or experimental allergic neuritis (Hughes and Colston, unpublished).

To my knowledge no other protection experiments have been carried out using cloned antigens other than the 65 kD protein. However recent results have suggested an alternative strategy for looking at the role of another *M. leprae* antigen, the 18 kD protein, in protective immunity. We have recently described a cultivable mycobacterium, *M. habana*, which is as effective as *M. leprae* itself in protecting mice against an *M. leprae* footpad infection (Singh *et al.*, 1989). On immunochemical analysis of *M. habana*, we have found that it reacts with a monoclonal antibody which was thought to be specific to *M. leprae*. This monoclonal antibody recognizes the 18 kD protein of *M. leprae* and we have confirmed the presence of both the protein and the gene in *M. habana* (Lamb *et al.*, 1990). This raises the question then of whether the 18 kD protein contributes to the ability of *M. habana* to protect against *M. leprae* infection. One approach to investigating this would be to follow that used for the 65 kD protein, that is to overproduce and purify the antigen and immunize mice with it. The problem with that approach is that we do not know the best way to deliver proteins in order to generate protective immunity. An alternative strategy which might be applicable in this case is to use a molecular genetic approach; since the 18 kD gene apparently has such a limited distribution, it might be reasonable to assume that it is a non-essential gene. If this is the case then it might be possible to use homologous recombination techniques to specifically delete the 18 kD gene from *M. leprae*. If this proves to be a possibility it would then be possible to compare the levels of protection conferred by bacteria which differ only at this locus.

5 THE WAY AHEAD

While the details of how mycobacteria are recognized by the cells of the immune system are becoming more clear, the understanding of how this recognition is translated into a protective immune response is becoming more confused. The role of cytotoxic T cells and of T cells expressing $\gamma \delta$ receptors is not clear; the activity of various lymphokines in activating macrophages to kill mycobacteria is controversial; and the role of specific groups of antigens in mediating protective immunity is not known. Even if effective stimulators of protective immunity can be identified, other problems will remain. One of these is the genetic restriction of the immune response to individual T-cell epitopes; it is well established that the T-cell response to proteins is restricted by the genetic background of the individual and thus any sub-unit vaccine will need to contain multiple T-cell sites each capable of inducing protection in individuals with different responder phenotypes. Another issue will be to find the most effective way of delivering proteins in order to generate long-term immunity. Speculation on the role of viral vectors (Moss, 1985) recombinant mycobacterial vaccines (Jacobs *et al.*, 1987) and a variety of adjuvants have been made, and such systems will require detailed evaluation. The ability to produce stable cosmid libraries in mycobacteria, and the availability of physical and genetic maps of mycobacterial genomes would permit the screening of a manageable number of recombinant mycobacteria such that the entire genome could be investigated.

In the meantime, a 'trial and error' approach using cloned antigens as they become available and using as wide a range of animal models and delivery systems as possible appears to be the only practical approach available.

REFERENCES

Billingham, M. E. J., Carney, S., Butler, R. and Colston, M. J. (1990). *J. Exp. Med.* **171**, 339–44.

Colston, M. J., and Hilson, G. R. F. (1976). *Nature* **262**, 339–401.

Fitch, F. W. (1986). *Microbiol. Rev.* **50**, 50–69.

Holoshitz, J., Koning, F., Coligan, J. E., De Bruyn, J. and Strober, S. (1989). *Nature* **339**, 226–229.

Jacobs, W. R., Tuckman, M. and Bloom, B. R. (1987). *Nature* **327**, 532–5.

Janis, E. M., Kaufman, S. H. E., Schwartz, R. H. and Pardoll, D. M. (1989). *Science* **244**, 713–16.

Kaufmann, S. H. E., Vath, U., Thole, J. E. R., Van Embden, J. D. A. and Emrich, F. (1987). *Eur. J. Immunol.* **17**, 351–7.

Kingston, A. E., Stagg, A. J. and Colston, M. J. (1986). *Immunology* **58**, 217–23

Lamb, F. I., Kingston, A. E., Estrada-G, I. and Colston, M. J. (1988). *Infect. Immun.* **56**, 1237–41.

Lamb, F. I., Singh, N. B. and Colston, M. J. (1990). *J. Immunol.* **144**, 1922–25.

Mehra, V., Brennan, P. J., Rada, E., Convit, J. and Bloom, B. R. (1984). *Nature* **308**, 194–6.

Modlin, R. L., Pirmez, C., Hofman, F. M., Torigian, V., Uyemura, K., Rea T. H., Bloom, B. R. and Brenner, M. B. (1989). *Nature* **339**, 544–8.

Moss, B. (1985). *Immunology Today* **6**, 243–5.

O'Brien, R. L., Happ, M. P., Dallas, A., Palmer, E., Kubo, R. and Born, W. K. (1989). *Cell* **57**, 667–74.

Orme, E. M. and Collins, F. M. (1986). *J. Exp. Med.* **163**, 203–8.

Ottenhoff, T. H. M., Klatser, P. R., Ivanyi, J., Elferink, D. G., De Wit, M. Y. L. and De Vries, R. R. P. (1986). *Nature* **319**, 66–8.

Ottenhoff, T. H. M., Birhane, K., Van Embden, J. D. A., Thole, J. E. R. and Kiessling, R. (1988). *J. Exp. Med.* **168**, 1947–52.

Shepard, C. C., Van Landingham, R. M. and Walker, L. L. (1980). *Infect. Immun.* **24**, 1034–39.

Shinnick, T. M. (1987). *J. Bacteriol.* **169**, 1080–8.

Singh, N. B., Lowe, A. C. R. E., Rees, R. J. W. and Colston, M. J. (1989). *Infect. Immun.* **57**, 653–5.

Thole, J. E. R., Keulen, W. J., Kolk, A. H. J., Groothuis, D. G., Berwald, L. G., Tiesjema, R. H. and Van Embden, J. D. A. (1987). *Infect. Immun.* **55**, 1466–1475.

4 Genome structure of mycobacteria

Josephine E. Clark-Curtiss

Departments of Molecular Microbiology and Biology, Washington University, St Louis, MO 63130, USA

1 INTRODUCTION

Many, if not all, mycobacteria are ubiquitous inhabitants of soil and water throughout the Earth; most are innocuous in terms of their relationships to humans and/or animals; many are beneficial to other living creatures in the ecosystem, due to their ability to fix nitrogen, to degrade organic material to forms usable for metabolism for other organisms, etc.; and a few mycobacterial species are pathogenic for humans and/or other animals (Ratledge and Stanford, 1982).

Historically, organisms were classified as mycobacteria on the basis of very slow growth rates in culture and on the property of acid-fast staining of the bacilli (Goodfellow and Wayne, 1982). Both of these criteria remain quite significant for classification, although the exact mechanisms responsible for these properties are not understood. Since the two major pathogenic species of the Genus *Mycobacterium* were identified (*M. leprae* in 1874 [Hansen] and *M. tuberculosis* in 1882 [Koch]), much has been learned about the chemical composition and structure of the complex mycobacterial cell walls (Barksdale and Kim, 1977; Goren and Brennan, 1979; Gaylord and Brennan, 1987; Brennan, 1989), about some metabolic capabilities (Ratledge, 1982; Barclay and Wheeler, 1989), about some antigenic determinants (Hunter *et al.*, 1982, 1986, 1989; Gaylord and Brennan, 1987) and a small amount has been learned about the pathogenic mechanisms of these bacilli (Ratledge and Stanford, 1983; Dannenberg, 1989; Ratledge *et al.*, 1989).

MOLECULAR BIOLOGY OF THE MYCOBACTERIA
ISBN 0–12–483378–0

Elucidation of the genetics of mycobacterial species has been greatly hampered by the slow growth characteristics of many species and the inability to cultivate *M. leprae* outside of animals (Grange, 1982). Initial studies dealt with determination of the guanine + cytosine (G+C) content of the DNA (Wayne and Gross, 1968; Bradley, 1973; Baess and Mansa, 1978; Imaeda *et al.*, 1982; Clark-Curtiss *et al.*, 1985) and determination of genome sizes (Bradley, 1973; Baess and Mansa, 1978; Baess, 1984; Clark-Curtiss *et al.*, 1985; McFadden *et al.*, 1987a). Table 4.1 summarizes these data for selected mycobacterial species. Most mycobacteria have between 64 and 70% G+C in their DNA; at present, there are only two exceptions to this grouping: *M. leprae* has a G+C content of 58% (Imaeda *et al.*, 1982; Clark-Curtiss *et al.*, 1985) and *M. lufu* has a G+C content of 61% (Clark-Curtiss *et al.*, 1985). Genome size determinations have revealed that most mycobacteria have large genomes (in the range of 3.1–4.5 x 10^9) compared to most other prokaryotes (Bradley, 1973; Baess and Mansa, 1978; McFadden *et al.*, 1987a). Interestingly, the pathogenic mycobacterial species have smaller genomes than other mycobacteria: *M. leprae* has a genome of 2.2 x 10^9 (Clark-Curtiss *et al.*, 1985), *M. tuberculosis* has a genome of 2.5 x 10^9 (Bradley, 1973) and *M. bovis* BCG has a genome of 2.8 x 10^9 (Bradley, 1973).

Table 4.1   Base Composition and Genome Sizes of Selected Mycobacterial DNAs.

Species	% Guanine + cytosine	Genome size (daltons)
A. Fast-growing species:		
M. fortuitum	65 (A)[a]	2.8 x 10^9 (B)
M. phlei	69 (A)	3.5 x 10^9 (B)
M. smegmatis	65–67 (A)[b]	4.2–4.5 x 10^9 (B)[b]
M. vaccae	65 (C)	3.1 x 10^9 (C)
B. Slow-growing species:		
M. bovis BCG	65 (A)	2.8 x 10^9 (B)
M. intracellulare	69 (A)	3.1 x 10^9 (B)
M. kansasii	64 (A)	4.2 x 10^9 (B)
M. lufu	61 (C)	3.1 x 10^9 (C)
M. marinum	65.5 (A)	3.8 x 10^9 (B)
M. paratuberculosis		
Strain Ben	66 (D)	3.1 x 10^9 (D)
M. tuberculosis H37Ra	65 (A)	2.5 x 10^9 (B)
M. leprae	58 (C)	2.2 x 10^9 (C)

[a] Data are from the following references: (A), Bradley, 1972; (B), Bradley, 1973; (C), Clark-Curtiss *et al.*, 1985; (D), McFadden *et al.*, 1987a.

[b] Ranges of values obtained with four different isolates of *M. smegmatis*.

Another early, indirect approach to studying genome structure was the use of DNA–DNA hybridizations among the genomes of mycobacterial species (Gross and Wayne, 1970; Bradley, 1973; Baess and WeisBentzon, 1978; Athwal et al., 1984; McFadden et al., 1987a; Grosskinsky et al., 1989). These experiments established that among certain groups of mycobacteria, significant amounts of the chromosomes must be similar in order that separated DNA strands from different species could anneal with one another when the denatured DNA was mixed and incubated together under appropriate conditions of salt concentration and temperature. Thus, it has been shown (summarized in Table 4.2) that DNA from members of the M. tuberculosis complex exhibited homologies between 78 and 97% to M. tuberculosis H37 chromosomal DNA and chromosomal DNAs from members of the fast-growing mycobacterial group were only 4 to 26% homologous to M. tuberculosis DNA (Athwal et al., 1984; Grosskinsky et al., 1989). M. paratuberculosis was shown to be very closely related (i.e. virtually indistinguishable) to M. avium strains by total chromosomal DNA–DNA hybridization experiments, whereas M. kansasii and M. phlei DNAs exhibited low degrees of homology to M. paratuberculosis DNA (McFadden et al., 1987a; Table 4.2B). Total chromosomal DNA–DNA hybridizations between M. leprae DNA and DNA from a number of cultivable mycobacteria (including both slow- and fast-growing species) showed that there was surprisingly little homology between M. leprae DNA and any other mycobacterial DNA (Athwal et al., 1984; Grosskinsky et al., 1989). An initial report of chromosomal DNA hybridizations between M. leprae DNA and DNAs from other mycobacteria and corynebacteria that appeared to show that M. leprae was more closely related to Corynebacterium than to mycobacteria (Imaeda et al., 1982) was later shown to be erroneous due to technical procedures used in the experiments (Athwal et al., 1984).

Chromosomal DNA-DNA hybridization experiments have established or confirmed related groups of mycobacteria by demonstrating amounts of homology among DNA sequences, but these experiments could not give information concerning the location of specific genes on the chromosomes to permit generation of genetic maps. In the absence of any identified systems for transfer of genetic information from one bacillus to another, other approaches have been necessary to begin to analyse specific genes and to correlate those genes with locations on their respective chromosomes.

2   IDENTIFICATION AND CHARACTERIZATION OF CONSERVED GENES OF MYCOBACTERIA

The development of recombinant DNA technology has been tremendously important in enabling researchers to begin genetic studies of organisms for

Table 4.2 Relatedness among Mycobacteria Determined by DNA–DNA Hybridizations.

Species	% Homology	% Homology
A. Relatedness to M. tuberculosis (Bradley, 1973)[a]:	H37Ra	Relatedness to M. tuberculosis (Erdmann) (Grosskinsky et al., 1989):
M. tuberculosis H37Ra	100	ND[b]
M. tuberculosis H37Rv	ND	95
M. tuberculosis (Erdmann)	ND	100
M. bovis BCG	86	97
M. intracellulare	29	26
M. kansasii	38	24
M. leprae	ND	2
M. lufu	ND	48
M. marinum	29	6
M. fortuitum	26	4
M. phlei	7	10
M. smegmatis	10–14[c]	4
M. vaccae	ND	<1
B. Relatedness to M. paratuberculosis Strain Ben (McFadden et al., 1987a)[a]:		
M. paratuberculosis Stran Ben	100	
M. paratuberculosis	100	
M. avium-intracellulare serovar 2	102	
M. avium-intracellulare serovar 5	107	
M. kansasii	32	
M. phlei	23	
C. Relatedness to M. leprae	(Grosskinsky et al., 1989)[a]:	(Athwal et al., 1984)[a]:
M. leprae (experimentally infected armadillo)	100	100
M. leprae (naturally infected armadillo)	ND	100
M. leprae (naturally infected Mangabey monkey)	ND	100
M. tuberculosis	8	21
M. bovis BCG	7	25
M. intracellulare	7	17
M. kansasii	9	13
M. lufu	12	ND
M. marinum	6	26
M. fortuitum	6	ND
M. phlei	10	ND
M. smegmatis	7	ND
M. vaccae	<1	13

[a] Data are from the references given in parentheses; per cent relatedness values are rounded off and standard deviations of the values are not presented, although they are given in the original papers.
[b] Not determined.
[c] Range of hybridization values obtained with four different isolates of M. smegmatis.

which classical genetic systems are not presently available and/or for organisms that are difficult to cultivate in the laboratory. Using recombinant DNA techniques, researchers have been able to identify and clone specific genes of *M. leprae*, *M. tuberculosis*, and *M. bovis* BCG. Simultaneously, other investigators have begun to define the chromosomes of these organisms and *M. avium* and *M. paratuberculosis* in molecular terms.

The low amounts of homology among mycobacterial chromosomes that were revealed by the chromosomal hybridization experiments are undoubtedly the result of hybridization between very highly conserved gene sequences. Such genes would be those that specify enzymes involved in intermediary metabolism, in the synthesis of characteristic mycobacterial structures such as phenolic glycolipids, lipoarabinomannans and mycolic acids, and genes that specify ribosomal RNA and proteins. A number of these genes have been identified and characterized: these include the genes for the common stress or heat-shock proteins (Chapter 1), the genes for the ribosomal RNA species and the gene for the enzyme citrate synthase from *M. leprae*.

2.1 Ribosomal RNA genes

Ribosomal RNA (rRNA) gene sequences are highly conserved among prokaryotes and have been used to establish phylogenetic relationships among numerous bacterial species (Woese *et al.*, 1983; Lane *et al.*, 1985). Bercovier *et al.* (1986) used a copy of the *Escherichia coli* rRNA operon as a probe in Southern (1975) hybridization experiments with chromosomal DNA from two fast-growing mycobacteria and four slow-growing mycobacteria. They determined that the fast-growing mycobacteria had two copies of the rRNA genes per chromosome, whereas the slow-growing mycobacteria had only a single copy of the genes (Bercovier *et al.*, 1986). Since other prokaryotes that are much faster growing than the mycobacteria have numerous copies of the rRNA genes (e.g. *E. coli* has seven copies; Brosius *et al.*, 1981), Bercovier *et al.* (1986) hypothesized that the low number of copies of rRNA genes in mycobacteria might be one of the reasons that the bacilli grow so much more slowly than other prokaryotes. These investigators also determined that the order of the rRNA genes in mycobacteria was –5′–16S–23S–5S–3′–, which is identical to the order of those genes in many other prokaryotes (Brosius *et al.*, 1981; Bercovier *et al.*, 1986, 1989). Later, Sela *et al.* (1989) used the *rrnB* operon of *E. coli* to determine that *M. leprae*, like the other slow-growing mycobacteria, also has only a single copy of the rRNA genes in its chromosome. However, Southern hybridizations using the *M. smegmatis* rRNA gene probe revealed that there were distinct differences in *Pst*I sites in the *M. leprae* rRNA genes compared to the gene sequences from other mycobacteria (Sela *et al.*, 1989).

Several groups have now determined the nucleotide sequences for 16S rRNA or the 16S rRNA gene from *M. avium, M. bovis, M. fortuitum, M. phlei, M. scrofulaceum, M. tuberculosis* H37Rv (Smida *et al.*, 1988); *M. bovis* BCG (Estrada-G. *et al.*, 1988; Suzuki *et al.*, 1988); and *M. leprae* (Smida *et al.*, 1988; Sela and Clark-Curtiss, 1990; Liesack and Stackebrandt, personal communication), either by sequence determination of reverse transcriptase-generated fragments of 16S rRNA (Estrada-G. *et al.*, 1988; Smida *et al.*, 1988), or by direct determination of the gene sequence (Suzuki *et al.*, 1988; Sela and Clark-Curtiss, 1990). These sequence determinations have revealed that the slow- and fast-growing species are well separated phylogenetically, but are all integral members of the actinomycete branch of the prokaryotic phylogenetic tree (Smida *et al.*, 1988). Among the slow-growing myco-bacterial species, an average of 96% of the sequences are homologous, indicating a tight cluster (Smida *et al.*, 1988). The *M. leprae* 16S rRNA sequence is approximately 95% homologous to the 16S rRNA sequences from *M. tuberculosis, M. bovis* and *M. avium*, indicating that it should be included among the group of slow-growing mycobacteria (Estrada-G. *et al.*, 1988; Smida *et al.*, 1988; Sela and Clark-Curtiss, 1990). However, certain regions of the *M. leprae* 16S rRNA exhibit significant structural variation from analogous regions in other mycobacterial 16S rRNA sequences (Estrada-G. *et al.*, 1988; Sela and Clark-Curtiss, 1990), reinforcing the idea that, although *M. leprae* is more like the slow-growing mycobacterial species than it is like other bacteria, this organism is distinctly different from other mycobacteria.

S. Sela has recently sequenced the 5' end of the 16S rRNA gene from *M. leprae* as well as regions upstream from the gene in order to study the promoter region for the gene. He has found that the promoter region sequences are very similar to consensus bacterial promoter sequences, both in actual nucleotide sequence and in the spacing of the -35 and -10 sequences relative to the beginning of the transcribed region (Sela and Clark-Curtiss, 1990). Moreover, he found a 24 nucleotide sequence between the promoter region and the coding sequence that is 75% homologous to sequences in similar locations in the 16S rRNA genes from the Gram-positive organisms *Bacillus subtilis* (Loughney *et al.*, 1983; Ogasawara *et al.*, 1983) and *Myco-plasma pneumoniae* (Gafney *et al.*, 1988; Taschke and Hermann, 1988); these sequences are involved in the processing of precursor 16S rRNA molecules into the mature forms (Sela and Clark-Curtiss, 1990).

2.2 Citrate synthase gene from *M. leprae*

The citrate synthase gene from *M. leprae* was the first mycobacterial gene to be identified, in terms of knowing the actual function of the gene (Jacobs *et al.*, 1986). This was done by identification of a clone from a pYA626::*M. leprae* DNA library that was able to complement a genetic defect in the biosynthesis of citrate synthase in a host *E. coli* strain. The *M. leprae* gene specified a protein of 46 kD (Jacobs *et al.*, 1986), which was very similar in size to citrate synthase of *E. coli* (Spencer and Guest, 1982; Ner *et al.*, 1983). When the *M. leprae* DNA insert fragment that specifies citrate synthase was used as a probe in Southern (1975) hybridization experiments with chromosomal DNAs from several mycobacterial species and *E. coli*, the probe hybridized to a fragment of all prokaryotic DNAs tested, but with varying degrees of intensity, indicating different amounts of sequence homology (Jacobs *et al.*, 1986; Clark-Curtiss and Walsh, 1989). Actual nucleotide sequence determinations of this gene are currently underway (N. Stoker, personal communication).

3   RESTRICTION FRAGMENT LENGTH POLYMORPHISM ANALYSES

An alternative approach to the identification, characterization and sequence determination of individual genes has been the use of a technique termed restriction fragment length polymorphism (RFLP) analysis (See McFadden *et al.*, this volume). This technique was developed in 1980 as a means to detect very small differences (i.e. single basepair changes) in chromosomes of organisms that are very closely related by conventional genomic analyses such as total chromosomal DNA–DNA hybridizations or sequence determinations of highly conserved genes (Botstein *et al.*, 1980). As the name implies, DNA restriction endonucleases are used to generate numerous fragments from chromosmal DNAs and the fragments are separated electrophoretically on the basis of size. Among closely related DNAs, differences in the length of a particular restriction fragment could be due to genotypic variations that resulted in differences of one or more individual bases such that cleavage sites for a given endonuclease could be lost or gained; alternatively, differences could be the result of insertions or deletions of blocks of DNA within the fragment (Botstein *et al.*, 1980). Using small, radioactively labelled restriction fragments as probe DNA's, variants among restriction fragments in total chromosomal DNA can be detected using the method of Southern hybridization (Southern, 1975). RFLP studies have been conducted among members of the *M. tuberculosis* complex (Eisenach *et al.*, 1986, 1988), the *M. paratuberculosis–M. avium* group (McFadden *et al.*, 1987b,c; 1988), and among isolates of *M. leprae* (Clark-Curtiss and Walsh, 1989; Williams and Gillis, 1989).

3.1   RFLP analyses among members of the *M. tuberculosis* complex

Initial RFLP experiments among the *M. tuberculosis* complex were done by Eisenach *et al.* (1986) in which the eight probes were randomly chosen clones from a λ1059::*M. tuberculosis* H37Rv genomic library. Chromosomal DNA was derived from three reference and two clinical isolates of *M. tuberculosis*, and from two reference strains each of *M. bovis* and *M. bovis* BCG. Four different restriction endonucleases (each with a 6 basepair recognition site) were used to digest the chromosomal DNAs. The digested fragments were separated by agarose gel electrophoresis, transferred to nitrocellulose filters and were used in Southern hybridizations with radio-actively labelled probe DNA (Southern, 1975). Eisenach *et al.*, (1986) found that two of the eight probes hybridized with homologous DNA fragments from all strains tested. Two other probes hybridized with some homologous fragments from all the strains, but also with other fragments that differed from strain to strain (i.e. these probes detected some polymorphisms). The remaining four probes hybridized to multiple restriction fragments of *M. tuberculosis* H37Rv DNA and to only one or two fragments of *M. tuberculosis* H37Ra, *M. bovis*, or *M. bovis* BCG DNAs. From these studies, Eisenach *et al.* (1986) concluded that some regions of the chromosomes of members of the *M. tuberculosis* complex were highly conserved, whereas the sequences have diverged in other regions. These experiments clearly showed that RLFP analysis could distinguish among isolates of *M. tuberculosis* and between *M. tuberculosis* and *M. bovis* (Eisenach *et al.*, 1986).

Since the probes used in their initial experiments were large fragments (approximately 15 kb) of *M. tuberculosis* DNA, Eisenach *et al.* (1988) conducted additional RFLP experiments in which they used smaller probes derived from an M13mp18::*M. tuberculosis* genomic library (probe fragments were 1.5 to 2.5 kb); these clones were chosen because they included some of the conserved regions of the *M. tuberculosis* chromosome. Among these probes were three fragments that hybridized only to members of the tuberculosis complex, but not to any of the other mycobacterial species that are commonly found in sputum specimens. Each of the three probes hybridized to multiple restriction fragments of chromosomal DNA from the tuberculosis complex members: many of the fragments to which the probes hybridized were homologous among the isolates tested but often the probes hybridized to additional fragments that were specific to a given isolate (Eisenach *et al.*, 1988). Thus, using these probes, Eisenach *et al.* (1988) were able to establish molecular 'fingerprints' for the different isolates which they tested and they speculated that such fingerprinting could be very useful for epidemiological studies of tuberculosis complex isolates.

3.2   RFLP analysis among members of the *M. avium–M. intracellulare–M. paratuberculosis* group

As described in Section 1, DNA–DNA hybridizations had shown that *M. paratuberculosis* DNA was indistinguishable from *M. avium* DNA (McFadden *et al.*, 1987a). However, *M. avium* is primarily a pathogen of birds that does not cause Johne's disease in ruminants; whereas *M. paratuberculosis* causes Johne's disease (regional enteritis) of ruminants but is avirulent for birds. To identify genetic differences between these pathogens, McFadden and his colleagues conducted RFLP experiments to attempt to distinguish between these mycobacteria. Using 10 independently derived random genomic clones obtained from a Crohn's disease-derived isolate (Ben) of *M. paratuberculosis* to probe DNA from *M. paratuberculosis* strains from Crohn's disease, the American type strain of *M. paratuberculosis*, *M. avium*, *M. kansasii* and the fast-growing mycobacterium, *M. phlei*. Of the 10 probes examined, three hybridized to all mycobacteria examined except *M. phlei*, four probes hybridized to all mycobacteria examined except *M. phlei* and *M. kansasii* and one probe (pMB22) hybridized to give multiple banding patterns with *M. paratuberculosis* and the Crohn's disease mycobacteria, but hybridized to far fewer bands with *M. avium* and weakly to the other mycobacteria. The degree of relationship of *M. paratuberculosis* to other mycobacteria was therefore: *M. paratuberculosis* > *M. avium* > > *M. kansasii* > *M. phlei*, the same as that suggested by DNA hybridization experiments (Table 4.2B). In addition to the result obtained for the clone pMB22, a number of RFLPs were detected that differentiated between *M. paratuberculosis* and *M. avium*. The base substitution, between *M. paratuberculosis* and *M. avium* was estimated for each of the probes independently as between 0 – 2.32%, with a mean at about 2%; indicating that the mutations detected by the RFLPs examined, were randomly distributed throughout the probe sequences and therefore the chromosomes of these mycobacteria. No RFLPs were found to distinguish between *M. paratuberculosis* strains; and the maximum frequency of base substitution between them was estimated as less than 0.15%, with 95% confidence limits. The results therefore indicated that although *M. paratuberculosis* and *M avium* were very closely related, *M. paratuberculosis* appeared to be a genetically distinct, highly conserved pathogen. These results have been confirmed by a study of a greater number of strains and serotypes of the *M. avium* complex and further *M. paratuberculosis* strains (McFadden *et al.*, 1987c and McFadden *et al.*, submitted). Since *M. avium* and *M. paratuberculosis* are so closely related but demonstrate distinct differences in pathogenicity, elucidation of the precise genetic changes that separate these pathogens may shed light on factors involved in mycobacterial virulence. The clone pMB22, which hybridized to multiple fragments of *M.*

paratuberculosis DNA is described further in Section 4 and by McFadden (this volume).

3.3   RFLP analyses among isolates of *M. leprae*

Although *M. leprae* shares very few nucleotide sequences with other myco-bacterial DNAs with which it has been tested, the few studies comparing chromosomes of *M. leprae* isolates by DNA–DNA hybridization indicated that the chromosomes were highly homologous (Athwal *et al.*, 1984; Meyers *et al.*, 1985). Moreover, since *M. leprae* cannot be cultivated on conventional mycobacteriological laboratory media, there has been no easy way to analyse isolates from different parts of the world to determine the extent of relatedness or divergence among *M. leprae* isolates. Thus, the technique of RFLP analysis seemed particularly appropriate for determining molecular relatedness among *M. leprae* isolates.

We initiated such analyses, using as probes a total of 15 clones from the pYA626::*M. leprae* genomic library (Clark-Curtiss *et al.*, 1985) and from the λgt11::*M. leprae* (Young *et al.*, 1985b). Four of the probes were *M. leprae* DNA fragments that contained identifiable genes (i.e. genes for citrate synthase, an enzyme that complemented a *purE* mutation in an *E. coli* host strain, and two antigenic determinants) and 11 probes were randomly chosen clones from the pYA626::*M. leprae* library (Clark-Curtiss *et al.*, 1985). Part of the *rrnB* operon from *E. coli* was also used, as a probe with a known functional gene (Bercovier *et al.*, 1986; Sela *et al.*, 1989). The chromosomal DNAs were from *M. leprae* isolates from India, West Africa and the southern US. In addition, the isolates were from human leprosy patients, a naturally infected Mangabey monkey (Meyers *et al.*, 1985) and a naturally infected armadillo (Walsh *et al.*, 1975). The chromosomal DNAs were digested with one restriction endonuclease that had a seven basepair recognition sequence, three endonucleases with six basepair recognition sequences and two endonucleases with four basepair recognition sequences (Clark-Curtiss and Walsh, 1989). The digested DNAs were treated and used in Southern hybridizations with the probes as described previously (Section 3.1 and Clark-Curtiss and Walsh, 1989).

As would be anticipated from the data obtained from the total chromosomal DNA–DNA hybridizations, most of the probes used for the RFLP experiments hybridized only to *M. leprae* DNA and not to DNA from other mycobacterial species. Of the 14 probes tested against DNA from cultivable mycobacteria, three probes hybridized to all mycobacterial DNAs tested, and one randomly chosen probe hybridized to *Mycobacterium lufu* DNA and to DNA from an armadillo-derived mycobacterial (ADM) strain (Portaels *et al.*, 1986) in addition to DNA from the *M. leprae* isolates tested (Clark-Curtiss and Walsh, 1989).

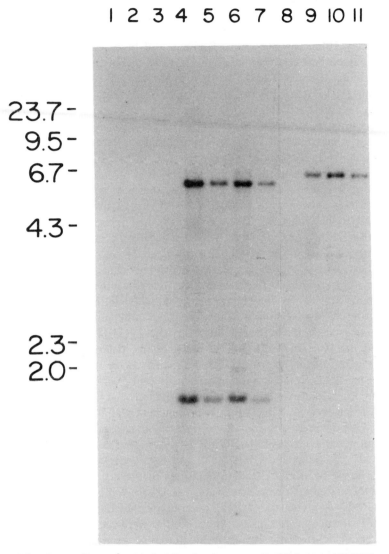

Figure 4.1 Autoradiograph of hybridization between the *M. leprae* pYA1070 probe DNA and restriction endonuclease-digested chromosomal DNAs. Lanes 2–7. chromosomal DNAs were digested with *Pst*I; Lanes 8–11: chromosomal DNAs were digested with *Bam*HI. Lane 1 — *Hind*III-digested bacteriophage λ DNA; Lane 2 — *E. coli* K-12 DNA; Lane 3 — uninfected armadillo DNA; Lane 4 — *M. leprae* H-1 DNA; Lanes 5 and 9 — *M. leprae* H-2 DNA; Lanes 6 and 10 — *M. leprae* A-1 DNA; Lanes 7 and 11 — *M. leprae* M-1 DNA; Lane 8 — uninfected human DNA. Reprinted from *J. Bacteriol.* **171**, 4844–51 (1989), with permission.

No RFLPs were found between the chromosomal DNAs of the *M. leprae* isolates from the human leprosy patients (which were from India). Using the same formula of Upholt (1977) that was used by McFadden *et al.* (1987b), we estimated that the maximum frequency of base substitution that could have occurred between these chromosomes would be less than 0.47%, with 95% confidence limits (Clark-Curtiss and Walsh, 1989).

In the RFLP experiments comparing DNA from the human Indian isolates, the West African Mangabey monkey isolate and the US armadillo isolate, a single polymorphism was detected (among 450 restriction fragments examined) that could distinguish among these isolates (Clark-Curtiss and Walsh, 1989). By combining the data from all of the RFLP experiments with these isolates and using the formula of Upholt (1977), we estimated that the maximum frequency of base substitution that has occurred among these chromosomes is between 0.02 and 0.26% (Clark-Curtiss and Walsh, 1989). An example of a typical RFLP experiment is depicted in Figure 4.1 in which the probe DNA (from pYA1070) hybridized to the same sized restriction fragments of chromosomal DNA from human *M. leprae* isolates, from the Mangabey monkey *M. leprae* isolate and from an armadillo *M. leprae* isolate when the chromosomal DNAs were digested with either *Pst*I or *Bam*HI.

Williams and Gillis (1989) have also conducted RFLP experiments with DNA from *M. leprae* isolates of diverse geographic and host origins. These investigators used insert fragments from the λgt11::*M. leprae* clones that specified the 70, 65, 28, 18 and 12 kD antigenic determinants of *M. leprae* (Young *et al.*, 1985b) as probes. Two of the *M. leprae* isolates were ones also used by us (the isolates from the naturally infected Mangabey monkey and the naturally infected armadillo); in addition, Williams and Gillis used five human isolates from Mexico, The Philippines, Thailand and the US. Chromosomal DNAs were digested with four restriction endonucleases that had six basepair recognition sequences and the RFLP analyses were conducted as described previously (Williams and Gillis, 1989 and Section 3.1). No polymorphisms were detected in any of the hybridizations of these probes to the chromosomal DNAs of the *M. leprae* isolates listed above (Williams and Gillis, 1989).

The RFLP analyses dramatically demonstrate the highly conserved nature of the *M. leprae* chromosomes. These results were quite unexpected, given the diverse origins of the *M. leprae* isolates: both geographical diversity and diversity of infected hosts.

3.4 Implications of the RFLP analyses among *M. leprae* isolates and among *M. paratuberculosis* isolates

The remarkable degree of conservation of nucleotide sequences among the isolates of *M. leprae* (Clark-Curtiss and Walsh, 1989; Williams and Gillis,

1989) and among the isolates of *M. paratuberculosis* (McFadden *et al.*, 1987b, 1988) raises some interesting questions about the nature of the chromosomal structure of these organisms. Since both *M. leprae* and *M. paratuberculosis* are very slow-growing bacteria, does this conservation of sequences imply or indicate that these organisms possess a special type of DNA polymerase, one that is either very faithful in its replication function or very specific in its repair function, or both? Do these bacteria have other, unique enzymes involved in repair of chromosomal mutations? Is the slow growth of these organisms due to the fact that the DNA is replicated slowly, perhaps because of extensive repairs that must be made at every replication cycle, since these intracellular organisms are certainly subjected to mutagenic substances in their environments?

Another consideration arising from the RFLP analyses has to do with the evolution of *M. leprae* and *M. paratuberculosis*. Although leprosy appears to be an ancient disease in terms of the history of mankind, *M. leprae* may be of recent origin when viewed from an evolutionary perspective. The organism we classify as *M. leprae* may have evolved from an existing progenitor soon after humans evolved; thus, all *M. leprae* in the world today may have descended from this single source. Since the organism occupies a very specific ecological niche, does not appear to have contact with many other bacterial species, and shares very few nucleotide sequences with other bacteria thus far tested, there may have been no opportunity for introduction of new genetic material into the *M. leprae* chromosome. However, even if one argues that *M. leprae* is of recent evolutionary origin, there still has been sufficient time for mutations to occur in its chromosomes, especially since the macrophage environment in which the bacilli grow contains numerous mutagenic substances. Thus, the argument for possession of very efficient replication and/or repair mechanisms becomes stronger.

4   REPETITIVE SEQUENCES IN MYCOBACTERIAL CHROMOSOMES

Within the last three years, there have been several reports of repeated DNA sequences in the chromosomes of *M. leprae* (Clark-Curtiss and Docherty, 1989; Grosskinsky *et al.*, 1989), *M. paratuberculosis* (McFadden *et al.*, 1988; 1989), and *M. tuberculosis* (Eisenach *et al.*, 1988; Reddi *et al.*, 1988, Zainuddin and Dale, 1989). Each of these sequences was initially identified by its hybridization patterns in RFLP analyses. Probes that contained the repetitive sequences hybridize to multiple restriction fragments of chromosomal DNA rather than to a single fragment as is the case with probes that include a unique fragment of the chromosome. This type of hybridization pattern is also seen when DNA fragment containing insertion sequences are used to probe genomic DNA (Dykhuizen *et al.*, 1985). DNA sequence data on some of the mycobacterial repetitive elements indicates that they also may be insertion sequences (see Martin *et al.*, this volume).

4.1   Repeated sequences in *M. tuberculosis* complex chromosomes

Eisenach *et al.* (1988) observed repetitive DNA sequences in *M. tuberculosis* DNA in their RFLP studies using the M13mp18::*M. tuberculosis* library described in Section 3.1. Three different probes were observed to hybridize to multiple bands of *M. tuberculosis* DNA. These probes were approximately 1.5–2 kb in size and hybridized to *M. tuberculosis* and *M. bovis* DNA, but produced different hybridization patterns for different isolates, thereby permitting Eisenach *et al.* (1988) to distinguish between *M. tuberculosis* and *M. bovis* and among different clinical isolates of *M. tuberculosis* by these patterns.

A second example of a clone bearing a repeated sequence of *M. tuberculosis* DNA was described by Reddi *et al.* (1988). This repeated sequence was on a 5.6 kb *Alu*I fragment isolated from a λgt11::*M. tuberculosis* H37Rv genomic library (Young *et al.*, 1985a). Reddi *et al.* (1988) demonstrated that the repetitive sequence present on this fragment hybridized to multiple *Alu*I fragments of DNA from *M. tuberculosis* H37Rv, *M. tuberculosis* H37Ra, and *M. bovis* BCG DNAs, but to only a single *Alu*I fragment of *M. kansasii* DNA. Hybridization with this probe also resulted in hybridization patterns that could distinguish between *M. tuberculosis* DNA and *M. bovis* BCG DNA (Reddi *et al.*, 1988).

Recently, Zainuddin and Dale (1989) have described a different repeated sequence present in *M. tuberculosis* DNA, which is homologous to a sequence present on a plasmid (pUS300) from *M. fortuitum*. This repetitive element hybridized to several homologous *Pvu*II fragments of DNA from 15 *M. tuberculosis* strains, such that Zainuddin and Dale could establish hybridization patterns characteristic for different isolates. Although the repetitive sequence probe also hybridized to DNA from *M. bovis* and *M. bovis* BCG, it hybridized to considerably fewer restriction fragments, thereby resulting in hybridization patterns which were quite different from those obtained with DNA from the *M. tuberculosis* isolates (Zainuddin and Dale, 1989; see McFadden, this volume).

DNA sequence data has demonstrated that the element described by Eisenach *et al.* (1988), is homologous to the *E. coli* insertion sequence, IS*3411* and has been designated IS*6110* (see Martin *et al.*, this volume). DNA sequence data of the element described by Zainuddin and Dale (1989) indicates that this element is almost identical to IS*6110* (McAdam *et al.*, unpublished).

4.2   Repeated sequences in the *M. paratuberculosis* genome

The probe pMB22, isolated by McFadden *et al.* (1987b) has already been mentioned (Section 2). The multiple banding patterns obtained with *M. paratuberculosis* and this clone were found to be due to the presence of a single copy of the insertion sequence IS*900*, in the clone (Green *et al.*, 1989; see Martin *et al.*, this volume). IS*900* is present in 10–15 copies in the genome of *M. paratuberculosis* but is absent in closely related *M. avium* strains. However, repetitive elements, related to IS*900* have been detected in some *M. avium* strains (McFadden *et al.*, this volume).

4.3   Repeated sequences in the *M. leprae* chromosome

In the course of the RFLP analyses comparing DNA from different isolates of *M. leprae*, one of the randomly chosen pYA626::*M. leprae* clones, designated pYA1065, was observed to hybridize to approximately 19 fragments of *Pst*I-digested chromosomal DNA from *M. leprae* (Clark-Curtiss and Docherty, 1989). As had been the case with the other probes used in RFLP analyses of *M. leprae* DNA, pYA1065 hybridized to exactly the same sized digestion fragments, regardless of the source of the *M. leprae* DNA. When other restriction endonucleases were used to digest the chromosomal DNAs, pYA1065 always hybridized to exactly the same sized fragments from all isolates tested, except for a single extra fragment that was present in the DNA of an *M. leprae* isolate from a naturally infected armadillo (Figure 4.2): this fragment was not present in the DNA of any of the other isolates tested (Clark-Curtiss and Docherty, 1989).

Moreover, this sequence appears to be specific to *M. leprae* DNA, since we did not observe hybridization of the pYA1065 insert fragment to DNA from *M. avium*, *M. bovis* BCG, *M. chelonei*, *M. fortuitum*, *M. intracellulare*, *M. lufu*, *M. tuberculosis*, or *M. vaccae* (Clark-Curtiss and Docherty, 1989). Thus, this probe is a good candidate for use as a probe for early diagnosis of leprosy. We were able to hybridize radioactively labelled pYA1065 probe DNA to DNA from approximately 10^5 disrupted *M. leprae* cells present in homogenized skin biopsy tissue from human leprosy patients (Clark-Curtiss and Docherty, 1989). Using the technique of polymerase chain reaction (PCR; Saiki *et al.*, 1985) to amplify target sequences in chromosomal DNA from *M. leprae* in biopsy samples, we and others (Woods and Cole, 1990) are attempting to increase the sensitivity of detection using the repetitive sequence probe.

The repeated sequence present in pYA1065 is present in 15 to 22 copies in the *M. leprae* chromosome. We have sequenced the *M. leprae* insert fragment from pYA1065 and have compared it to the published sequence of another

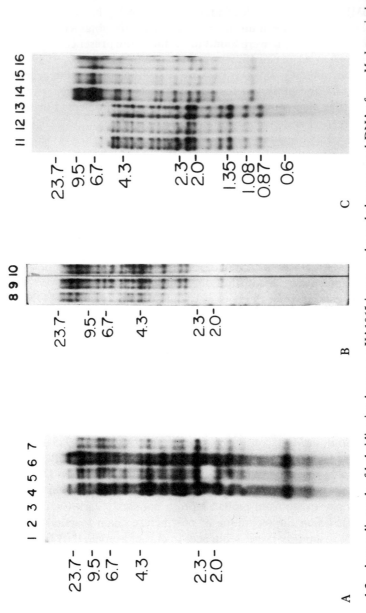

Figure 4.2 Autoradiograph of hybridization between pYA1065 insert probe and chromosomal DNAs from *M. leprae* isolates. Panel A: Chromosomal DNAs were digested with *Pst*I. Panel B: Chromosomal DNAs were digested with *Bam*HI. Panel C: Chromosomal DNAs were digested with *Bst*EII (Lanes 11–13) and *Sac*II (Lanes 14–16). Lane 1 — *Hind*III-digested bacteriophage λDNA; Lane 2 — uninfected armadillo; Lane 3 — *E. coli* K-12; Lane 4 — human *M. leprae* No. 29; Lanes 5, 8, 11, and 14 — human *M. leprae* No. 93; Lanes 6, 9, 12, and 15 — *M. leprae* from naturally infected armadillo; Lanes 7, 10, 13, and 16 — *M. leprae* from naturally infected Mangabey monkey. Reprinted from *J. Infectious Diseases* **159**, 7–15 (1989), with permission.

example of the repeated sequence (Grosskinsky *et al.*, 1989), which is located downstream from the coding region for the 65 kD antigenic determinant/ stress protein of *M. leprae* in Y3178 (Mchra *et al.*, 1986). The repeated sequence is 600 bp long and although there are portions which are able to form stem-loop structures and other portions that form small or imperfect indirect repeats, at the present time, we do not know the function of the repeated sequence. We do not have evidence that any protein is actually specified and produced by the *M. leprae* repeated sequence, either by minicell experiments (Roozen *et al.*, 1971) or by *in vitro* transcription–translation experiments (Galan and Curtiss, 1989). We are presently characterizing additional clones that contain the repeated sequence in an effort to determine whether or not the repeated sequence is located near a particular class of genes, or whether the repeated sequence may be involved in maintaining the topography of the chromosome.

5 CONCLUSIONS

In the past decade, tremendous strides have been made in our understanding of mycobacterial chromosomes, yet we are just beginning to learn about specific genes, about features such as repetitive sequences, about the remarkable conservation of sequences among the most slow-growing myco- bacteria. In the next few years, we will be determining the sites of specific genes on chromosomal fragments, either by pulse-field electrophoretic studies (Carle *et al.*, 1986) or by production of overlapping cosmid libraries (Wenzel and Hermann, 1988). As more genes from different mycobacterial species are identified and sequenced and as their upstream regulatory sequences are studied, we will begin to understand more about the regulation of transcription in these organisms. In conjunction with the mycobacterial cloning systems being developed by W. R. Jacobs and his colleagues (described in Chapter 9), we should be able to study mechanisms of translation in mycobacteria and make even greater advances in our under- standing of both the genetics and physiology of these very important and interesting bacteria.

ACKNOWLEDGMENTS

Research discussed in this presentation from the Clark-Curtiss group was supported by US Public Health Service grants AI-23470 and AI-26186 from the National Institutes of Allergy and Infectious Diseases.

REFERENCES

Athwal, R. S., Deo, S. S. and Imaeda, T. (1984). *Int. J. Syst. Bacteriol.* **34**, 371–5.

Baess, I. (1984). *Acta Pathol. Microbiol. Immunol. Scand., Sect. B* **92**, 209–11.

Baess, I. and Mansa, B. (1978). *Acta Pathol. Microbiol. Scand., Sect. B* **86**, 309–12.

Baess, I. and WeisBentzon, M. (1978). *Acta Pathol. Microbiol. Scand., Sect. B* **86**, 71–6.

Barclay, R. and Wheeler, P. R. (1989). In *The Biology of the Mycobacteria*, Vol. 3 (eds C. Ratledge, J. Stanford and J. Grange), pp.37–106. Academic Press, London.

Barksdale, L. and Kim, K. S. (1977). *Bacteriol. Rev.* **41**, 217–372.

Bercovier, H., Kafri, O. and Sela, S. (1986). *Biochem. Biophys. Res. Commun.* **136**, 1136–41.

Bercovier, H., Kafri, O., Kornitzer, D. and Sela, S. (1989). *FEMS Microbiol. Lett.* **57**, 125–8.

Botstein, D., White, R. L., Skolnick, M. and Davis, R. W. (1980). *Am. J. Hum. Genet.* **32**, 314–31.

Bradley, S. G. (1972). *Adv. Front. Plant Sci.* **28**, 349–62.

Bradley, S. G. (1973). *J. Bacteriol.* **113**, 645–51.

Brennan, P. J. (1989). *Rev. Inf. Dis.* **11** (Suppl. 2), S420–30.

Brosius, J., Ullrich, A., Raker, M. A., Gray, A., Dull, T. J., Gutell, R. R. and Noller, H. F. (1981). *Plasmid* **6**, 112–18.

Carle, G. F., Frank, M. and Olson, M. V. (1986). *Science* **232**, 65–8.

Clark-Curtiss, J. E. and Docherty, M. A. (1989). *J. Inf. Dis.* **159**, 7–15.

Clark-Curtiss, J. E. and Walsh, G. P. (1989). *J. Bacteriol.* **171**, 4844–51.

Clark-Curtiss, J. E., Jacobs, W. R., Jr, Docherty, M. A., Ritchie, L. R. and Curtiss, R. III. (1985). *J. Bacteriol.* **161**, 1093–102.

Dannenberg, A. M., Jr (1989). *Rev. Inf. Dis.* **11** (Suppl. 2), S369–78.

Dykhuizen, D. E., Sawyer, S. A., Green, L., Miller, R. D. and Hartl, D. L. (1985). *Genetics* **111**, 219–31.

Eisenach, K. D., Crawford, J. T. and Bates, J. H. (1986). *Am. Rev. Resp. Dis.* **133**, 1065–8.

Eisenach, K. D., Crawford, J. T. and Bates, J. H. (1988). *J. Clin. Microbiol.* **26**, 2240–5.

Estrada-G., I. C. E., Lamb, F. I., Colston, M. J. and Cox, R. A. (1988). *J. Gen. Microbiol.* **134**, 1449–53.

Gafney, R., Hyman, H. C. and Glaser, G. (1988). *Nucl. Acids Res.* **16**, 61–76.

Galan, J. E. and Curtiss, R. III. (1989). *Proc. Natl. Acad. Sci. USA* **86**, 6383–7.

Gaylord, H. and Brennan, P. J. (1987). *Ann. Rev. Microbiol.* **41**, 645–75.

Goodfellow, M. and Wayne, L. G. (1982). In *The Biology of the Mycobacteria*, Vol. 1 (eds C. Ratledge and J. Stanford) pp.471–521. Academic Press, London.

Goren, M. B. and Brennan, P. J. (1979). In *Tuberculosis* (ed. G. P. Youmans), pp. 63–193. W. B. Saunders, Philadelphia.

Grange, J. M. (1982). In *The Biology of the Mycobacteria*, Vol. I (eds C. Ratledge and J. Stanford), pp. 309–51. Academic Press, London.

Green, E. P., Tizard, M. L. V., Moss, M. T., Thompson, J., Winterbourne, D. J., McFadden, J. J. and Hermon-Taylor, J. (1989). *Nucleic Acids Res.* **17**, 9063–73.

Gross, W. M. and Wayne, L. G. (1970). *J. Bacteriol.* **104**, 630–4.

Grosskinsky, C. M., Jacobs, W. R., Clark-Curtiss, J. E. and Bloom, B. R. (1989). *Infect. Immun.* **57**, 1535–41.

Hansen, G. A. (1874). *Norsk. Mag. Laegervidenskaben* **4**, 76–9.

Hunter, S. E., Fujiwara, T. and Brennan, P. J. (1982). *J. Biol. Chem.* **257**, 15072–8.

Hunter, S. W., Gaylord, H. and Brennan, P. J. (1986). *J. Biol. Chem.* **261**, 12345–51.

Hunter, S. W., McNeil, M., Modlin, R. L., Mehra, V., Bloom, B. R. and Brennan, P. J. (1989). *J. Immunol.* **142**, 2864–72.

Imaeda, T., Kirchheimer, W. F. and Barksdale, L. (1982). *J. Bacteriol.* **150**, 414–17.

Jacobs, W. R., Docherty, M. A., Curtiss, R. III and Clark-Curtiss, J. E. (1986). *Proc. Natl. Acad. Sci. USA* **83**, 1926–30.

Koch, R. (1882). *Berlin. Klin. Wochenschr.* **19**, 221–30.

Lane, D. J., Pace, B., Olsen, G. J., Stahl, D. A., Sogin, M. L. and Pace, N. R. (1985). *Proc. Natl. Acad. Sci. USA* **82**, 6955–9.

Loughney, K., Lund, E. and Dahlberg, J. E. (1983). *Nucl. Acids Res.* **11**, 6709–21.

McFadden, J. J., Butcher, P. D., Chiodini, R. J. and Hermon-Taylor, J. (1987a). *J. Gen. Microbiol.* **133**, 211–14.

McFadden, J. J., Butcher, P. D., Chiodini, R. and Hermon-Taylor, J. (1987b). *J. Clin. Microbiol.* **25**, 796–801.

McFadden, J. J., Butcher, P. D., Thompson, J., Chiodini, R. and Hermon-Taylor, J. (1987c). *Molecular Microbiology* **1**, 283–91.

McFadden, J. J., Thompson, J., Hull, E., Hampson, S., Stanford, J. and Hermon-Taylor, J. (1988). In *Inflammatory Bowel Disease: Current Status and Future Approach*, (ed. R. P. MacDermott), pp. 515–20. Elsevier Science Publishers, London.

McFadden, J. J., Green, E. P., Thompson, J., Moss, M., Tizard, M. L. V., Hampson, S., Portaels, F. and Hermon-Taylor, J. (1990). Submitted.

Mehra, V., Sweetser, D. and Young, R. A. (1986). *Proc. Natl. Acad. Sci. USA* **83**, 7013–17.

Meyers, W. M., Walsh, G. P., Brown, H. L., Binford, C. H., Imes, G. D., Jr, Hadfield, T. L., Schlagel, C. J., Fukunishi, Y., Gerone, P. J., Wolf, R. H., Gormus, B. J., Martin, L. N., Harboe, M. and Imaeda, T. (1985). *Int. J. Lepr.* **53**, 1–14.

Ner, S. S., Bhayana, V., Bell, A. W., Giles, I. G., Duckworth, H. W. and Bloxam, D. (1983). *Biochemistry* **22**, 5243–8.

Ogasawara, N., Moryia, S. and Yoshikawa, H. (1983). *Nucl. Acids Res.* **11**, 6301–18.

Portaels, F., Asselineau, C., Baess, I, Daffee, M., Dobson, G., Draper, P., Gregory, D., Hall, R. M., Imaeda, T., Jenkins, P. A., Laneelle, M. A., Larsson, L., Magnussen, M., Minnikin, D. E., Pattyn, S. R., Wieten, G. and Wheeler, P. R. (1986). *J. Gen. Microbiol.* **132**, 2693–707.

Ratledge, C. (1982). In *The Biology of the Mycobacteria*, Vol. 1 (eds C. Ratledge and J. Stanford), pp. 186–271. Academic Press, London.

Ratledge, C. and Stanford, J. (Eds) (1982). *The Biology of the Mycobacteria*, Vol. 1. Academic Press, London.

Ratledge, C. and Stanford, J. (Eds) (1983). *The Biology of the Mycobacteria*, Vol. 2. Academic Press, London.

Ratledge, C., Stanford, J. and Grange, J. (Eds) (1989). *The Biology of the Mycobacteria*, Vol. 3. Academic Press, London.

Reddi, P. P., Talwar, G. P. and Khandekar, P. S. (1988) *Int. J. Lepr.* **56**, 592–8.

Roozen, K. J., Fenwick, R. G., Jr, and Curtiss, R. III. (1971). *J. Bacteriol.* **107**, 21–33.

Saiki, R. K., Scharf, S., Faloona, F. A., Mullis, K. B., Horn, G. T., Ehrlich, H. A. and Arnheim, N. (1985). *Science* **230**, 1350–4.

Sela, S. and Clark-Curtiss, J. E. (1990). Submitted to *J. Bacteriol.*

Sela, S., Clark-Curtiss, J. E. and Bercovier, H. (1989). *J. Bacteriol.* **171**, 70–3.

Smida, J., Kazda, J. and Stackebrandt, E. (1988). *Int. J. Lepr.* **56**, 449–53.

Spencer, M. E. and Guest, J. R. (1982). *J. Bacteriol.* **151**, 542–52.

Southern, E. M. (1975). *J. Mol. Biol.* **98**, 503–17.

Suzuki, Y., Nagata, A., Ono, Y. and Yamada, T. (1988). *J. Bacteriol.* **170**, 2886–9.

Taschka, C. and Hermann, R. (1988). *Mol. Gen. Genet.* **212**, 522–30.

Upholt, W. B. (1977). *Nuc. Acids Res.* **4**, 1257–65.

Walsh, G. P., Storrs, E. E., Burchfield, H. P., Meyers, W. M. and Binford, C. H. (1975). *J. Reticuloendothel. Soc.* **18**, 347–51.

Wayne, L. G. and Gross, W. M. (1968). Base composition of deoxyribonucleic acid isolated from mycobacteria. *J. Bacteriol.* **96**, 1915–19.

Wenzel, R. and Hermann, R. (1988). *Nuc. Acids Res.* **16**, 8323–36.

Williams, D. L. and Gillis, T. P. (1989). *Acta Leprologica* 7, Suppl. 1, 5226–30.

Woese, C. R., Gutell, R., Gupta, R. and Noller, H. (1983). *Microbiol. Rev.* **47**, 621–69.

Woods, S. A. and Cole, S. T. (1990). *FEMS Microbiol. Lett.* **65**, 305–310.

Young, R. A., Bloom, B. R., Grosskinsky, C. M., Ivanyi, J., Thomas, D. and Davis, R. W. (1985a). *Proc. Natl. Acad. Sci. USA* **82**, 2583–7.

Young, R. A., Mehra, V., Sweetser, D., Buchanan, T., Clark-Curtiss, J. E. and Bloom, B. R. (1985b). Genes for the major protein antigens of the leprosy parasite *Mycobacterium leprae*. *Nature (London)* **316**, 450–2.

Zainuddin, Z. F. and Dale, J. W. (1989). *J. Gen. Microbiol.* **135**, 2347–55.

5   Plasmids of the *Mycobacterium avium* complex

Jack T. Crawford and Joseph O. Falkinham, III*

Medical Research Service, John L. McClellan Memorial Veterans Hospital, and the Department of Microbiology and Immunology, University of Arkansas for Medical Sciences, 4300 W 7th, Little Rock, AR 72205, USA
* Department of Biology, Virginia Polytechnic Institute and State University, Blacksburg, VA 24061, USA

1   INTRODUCTION

Mycobacterial plasmids were first demonstrated in strains of the *Mycobacterium avium* complex (Crawford and Bates, 1979; Crawford *et al.*, 1981a) and subsequently plasmids have been shown to be common in this group and in the *Mycobacterium fortuitum* complex. The presence of plasmids in mycobacteria is significant for several reasons. In other bacteria, plasmids frequently encode drug resistance, virulence factors, antigens, and various functions that promote survival in the environment. Plasmids are excellent markers for epidemiological studies since they can be physically characterized and fingerprinted by restriction digestion and hybridization. Lastly, plasmids are tools for the development of vectors for molecular cloning and will greatly assist genetic analysis of the mycobacteria.

The nomenclature of the *M. avium* complex (MAC) is unsettled, but in the context of plasmid analysis it is clear that similar plasmids are found in strains referred to as *M. avium*, *M. intracellulare*, and *M. scrofulaceum*. We will use the term MAIS to refer to this group. The term MAC will be reserved for groups of clinical isolates that do not normally include *M. scrofulaceum*. Evidence of the increasing incidence of human infections caused by members

MOLECULAR BIOLOGY OF THE MYCOBACTERIA
ISBN 0–12–483378–0

of the MAIS group in persons with acquired immunodeficiency syndrome (AIDS) (Greene *et al.*, 1982; Blaser and Cohn, 1986) and non-AIDS patients (Good, 1980; Good and Snider, 1982) and the widespread distribution of members of that group in southeastern waters (Falkinham *et al.*, 1980), aerosols (Wendt *et al.*, 1980) and soils (Brooks *et al.*, 1984) has stimulated interest in their epidemiology, ecology, physiology, and genetics. Unlike *M. tuberculosis* which is highly evolved for survival in the infected host, the MAIS group is very heterogeneous with at least some members capable of proliferating in the environment (George *et al.*, 1980) and in tissues of animals and humans.

One important determinant of the ability of microorganisms to survive and proliferate in a variety of habitats is plasmid-encoded gene products. Members of the MAIS group carry a variety of plasmids, and those strains with plasmids often contain several (average 2.5–3.3/strain; Meissner and Falkinham, 1986). Though procedures for the reproducible isolation of plasmids from MAIS strains in amounts sufficient for restriction endo-nuclease digestion have been developed and a number of plasmid-encoded functions have been identified, relatively little is known about the role of plasmids in the epidemiology, ecology, and physiology of members of this group. Specific genetic studies have lagged because of a lack of easy means for genetic manipulation of these slow-growing and potentially pathogenic organisms. The introduction of new technology, especially the development of techniques for DNA transfection by electroporation and the identification of suitable selectable markers (W. R. Jacobs, this volume), will change this situation.

2   ISOLATION OF PLASMIDS

The isolation of covalently closed circular plasmid DNA requires gentle lysis of the cells. Intact plasmids, especially large plasmids, cannot be obtained by physical techniques such as sonication, passage through a French pressure cell, or shaking with glass beads. In addition, mycobacteria are fairly resistant to digestion with lysozyme or proteases, and are not readily lysed by detergent treatment. We adapted the previously reported technique of growth of mycobacteria in the presence of D-cycloserine or glycine to render the cells susceptible to lysis with detergents (Mizuguchi and Tokunaga, 1970). D-cycloserine interferes with cell-wall synthesis and, therefore, treat-ment requires active growth of the culture. With MAIS strains best growth is obtained with agitated cultures in a medium containing oleic acid. A modified 7H9 medium (designated 7H9-Av) containing 0.5% glycerol and Dubois oleic albumin enrichment in place of the usual Middlebrook ADC enrichment provides very dense growth (Crawford *et al.*, 1981c). Actively growing cells are treated with D-cycloserine at a final concentration of 1 mg/ml. Although MAIS strains are routinely resistant to cycloserine at therapeutic levels, they are susceptible to this very high concentration. Glycine can be substituted but it is not nearly as effective. Cultures are incubated until there is a slight drop in turbidity, usually 12–18 hours. Unlike other mycobacteria, MAIS strains do not produce significant levels of beta-lactamase, and lysis can be enhanced by addition of ampicillin at a final concentration of 100 μg/ml along with D-cycloserine. The major problem in obtaining good preparations of MAC plasmid DNA is obtaining efficient lysis of cells. It is important that the D-cycloserine and ampicillin are added while the cells are still in logarithmic phase and before stationary phase of growth. They must be added during mid log phase to ensure lysis of the majority of cells.

Initially the salt-cleared lysate procedure (Guerry, *et al.*, 1973) in conjunction with CsCl-ethidium bromide gradient centrifugation was used for the isolation of plasmid DNA (Crawford and Bates, 1979). Large volumes of culture (400–1000 ml) were used, and the cells were suspended in a small volume for lysis. In this procedure cells are lysed with sodium dodecyl sulphate (SDS), and cell debris and high molecular weight chromosomal DNA are precipitated with NaCl in the cold. Plasmid DNA is isolated as a satellite band on the caesium gradient. Our best preparations are obtained using a large capacity fixed angle rotor. The satellite band containing plasmid DNA is removed, mixed with 50% CsCl, and centrifuged in a swinging bucket rotor to concentrate the DNA. This procedure yields large amounts of very pure plasmid DNA, but it is not useful for preparing DNA from large numbers of strains.

Subsequently, we adapted the alkaline lysis procedure of Kado and Liu (1981) for routine screening of strains. In this procedure, SDS and high pH are used to lyse the cells and denature the DNA. Lysates are heated to degrade RNA and chromosomal DNA. Lysates are then extracted with unbuffered phenol-chloroform which lowers the pH, allowing renaturation of the plasmid DNA and extracts the protein and denatured high molecular weight chromosomal DNA. This procedure is particularly useful for mycobacteria since the high pH and heat treatment enhance lysis. In Little Rock we routinely apply this procedure to small cultures (10 ml cultures in flasks or 2–3 ml cultures in tubes incubated on a rotator) and often carry out the lysis and extraction in microcentrifuge tubes. Treatment with D-cycloserine is required. Cells are suspended in E-buffer at 1/20 volume of the original culture. The lysis procedure is carried out as described by Kado and Liu (1981) using a heat treatment 55–60°C for 20 minutes. High quality redistilled phenol is required for the extraction step. Although the samples

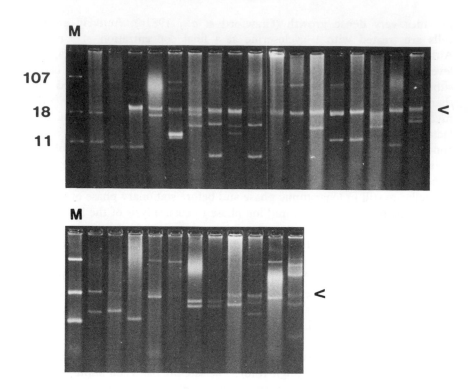

Figure 5.1 Agarose gel electrophoresis of plasmid DNA preparations from MAIS strains. DNA was prepared by the procedure of Kado and Liu (1981). Marker (M) is plasmid DNA from strain LR25 prepared by caesium chloride–ethidium bromide centrifugation, with the sizes of the three plasmids indicated in Mdal. Residual chromosomal DNA is indicated by the arrow. The lanes correspond to the strains listed in Table 5.1, strains 1–17 on the top gel and 18–28 on the bottom gel.

can be stored for several days, best results are obtained if the samples are electrophoresed immediately after extraction. Good results are obtained with 20–50 μl aliquots of lysate run on 20 × 20 cm gels. Minigels can also be used. The gels are run without ethidium bromide and stained after electrophoresis. Examples of excellent, but not necessarily typical, preparations are shown in Figure 5.1. The location of residual chromosomal DNA is indicated. Heating at higher temperatures or for longer periods reduces the distinct band of chromosomal DNA but produces smears that can obscure the plasmid bands. Mobility of the plasmids relative to the chromosomal DNA varies with the electrophoresis buffer and conditions. Figure 5.1 illustrates the common pattern seen with MAIS strains — small plasmids of 8.8–25 Mdal (13.5–38 kb) and large plasmids greater than 100 Mdal. (150 kb). To avoid confusion, the designation of size in Mdal is carried over from the original publications, when appropriate. The calculated equivalent in kilobase pairs is indicated.

Table 5.1 Plasmid Containing Strains.

No.[1]	Strain	Serotype	No.	Strain	Serotype
1	LR156	2	15	LR129	41
2	LR113	4	16	TMC1321	43
3	LR150	8	17	LR122	43
4	LR120	13	18	LR138	27
5	LR140	6	19	LR145	18
6	LR144	16	20	LR154	12
7	LR127	19	21	LR119	13
8	LR135	22	22	TMC1324	U
9	TMC1469	19	23	LR231	2
10	TMC1306	9	24	LR232	2
11	TMC1307	U[2]	25	LR264	ND[3]
12	TMC1312	41	26	LR256	ND
13	TMC1316	43	27	LR205	ND
14	TMC1314	41	28	LR206	ND

[1] No. refers to lanes in Figure 5.1.
[2] U, untypable.
[3] ND, not determined.

Although this procedure works well for screening strains for plasmids, it is our experience that plasmid DNA prepared by this method cannot be subsequently purified for use in molecular biology techniques as described by Kado and Liu. Extraction with ether, dialysis, etc. invariably result in a cloudy precipitate and loss of the DNA. Small plasmids can be recovered from these crude preparations by binding to powdered glass using the GeneClean reagent kit (BIO 101, Inc., LaJolla, CA; M. T. Jucker and L. Via; T. Hellyer, personal communication). However, large plasmids are broken in this procedure. Alternatively, individual plasmids can be recovered by excision from low-melting agarose gels.

Plasmid DNA suitable for restriction analysis can be prepared from large cultures by a modification of the Kado and Liu procedure (Meissner and Falkinham, 1984; Jucker *et al.*, in preparation). After cycloserine–ampicillin treatment, the cells are harvested by centrifugation and suspended in E-buffer to 1/40 volume of original culture. Cells are lysed by addition of two volumes of the alkaline-SDS lysing solution, and the sample is then heated at 60°C for 20 minutes. The lysate is extracted with phenol saturated with 0.5 M NaCl rather than the phenol-chloroform specified in the original procedure. This extraction is carried out in a roller bottle three-times the volume of the sample using slow rotation (4 rpm) for at least 30 minutes. Recovery of high molecular weight plasmid DNA (>150 kb) and removal of chromosomal DNA is influenced strongly by phenol extraction. If the cell lysate is mixed rapidly by inversion, rolling, or vortexing, large plasmids are not recovered and the preparation contains only small plasmids and a large amount of undenatured chromosomal DNA. Following centrifugation, the aqueous phase is gently extracted with phenol-chloroform-isoamyl alcohol. The DNA is then precipitated with ethanol. In some instances, ethanol precipitation yields a large amount of material which gives cloudy lanes in agarose gels and contains an inhibitor of restriction endonuclease activity. If this occurs, ammonium acetate extraction (Maniatis *et al.*, 1982) followed by ethanol precipitation yields solutions of plasmid DNA which are suitable for restriction endonuclease digestion. Addition of ammonium acetate to a final concentration of 2.5 M before ethanol precipitation results in precipitation of a large amount of fibrous material which does not contain plasmid DNA. The DNA is precipitated by ethanol, and it is best to suspend the DNA-containing pellet in a volume of at least half the volume of the original cell lysate. If the pellet is suspended in a smaller volume, up to 90% of the DNA can be lost in a subsequent ammonium acetate precipitation step.

One variable influencing plasmid isolation is colonial type. Opaque, domed colony variants yield more DNA than do transparent variants, possibly because the transparent cells are more resistant to lysis. The pigmented, opaque MAC isolates typically recovered from AIDS patients (Kiehn *et al.*, 1985) also yield more plasmid DNA than do their unpigmented, flat, transparent variants (Stormer, Jucker, and Falkinham, unpublished). This may also be due to resistance of the unpigmented variants to antibiotic-enhanced lysis, since these unpigmented colonial variants are more antibiotic resistant than their pigmented parents (Stormer and Falkinham, 1989).

The small plasmids can also be isolated by the standard alkaline extraction procedure used for *E. coli* (Birnboim and Doly, 1979). Cultures must be treated with D-cycloserine before processing, and incubation with lysozyme aids lysis. Lysis is also enhanced by incubation at 37–50°C after addition of the alkaline-SDS lysis buffer. Plasmids prepared in this manner are suitable for restriction digestion However, a high percentage of the plasmids are nicked in this procedure, and it should not be used for initial screening and size determinations. It has been our experience that repeated cycles of plasmid purification result in the appearance of topoisomers. Thus, multiple plasmid bands in a single strain should be interpreted with caution.

3   PHYSICAL CHARACTERIZATION OF SPECIFIC PLASMIDS

As discussed above, the majority of plasmids in MAIS strains fall into two size classes; small plasmids of 13.5–38 kb and large plasmids of >150 kb. One well-characterized strain which shows this pattern is *M. avium* LR25 (serotype 6). The sizes of the three plasmids of this strain were determined by electron microscopy to be 11.2, 18.3, and 107 Mdal (17.2, 28, and 165 kb) (Figure 5.2; Crawford *et al.*, 1981b). There is one report of a 2.2 Mdal (3.4 kb) plasmid (Mizuguchi *et al.*, 1981) and another of a 4.5 Mdal (6.9 kb) plasmid (Masaki *et al.*, 1989). Intermediate plasmids of 40–60 Mdal (60–90 kb) are seen in some strains. One strain with a plasmid of 350 Mdal (540 kb) was isolated (Meissner and Falkinham, 1986). It is not uncommon to find strains with at least 30% of the cell's total DNA represented as distinct plasmids.

3.1   pLR7

The most extensively studied *M. avium* plasmid is pLR7 (15.3 kb). This plasmid is carried by strain LR113 (serotype 4) and is the only plasmid

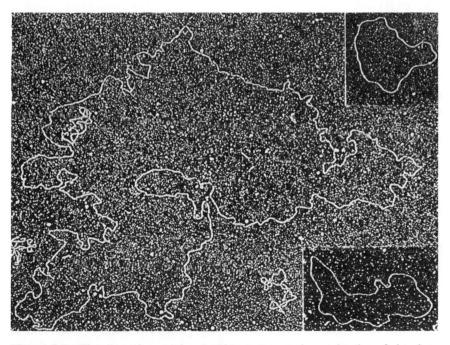

Figure 5.2 Electron micrographs showing representative molecules of the three plasmids from strain LR25.

carried by this strain (Table 5.2). Segments of LR7 were cloned into *E. coli* and the plasmid was physically mapped by restriction endonuclease digestion (Figure 5.3; Crawford and Bates, 1984). One clone, pJC20, consists of the complete pLR7 sequence joined to pBR322 via the unique HindIII site. This clone has been used in several studies as a hybridization probe to detect related plasmids. Unlike fragments of other bacterial DNA, large (>15 kb) cloned fragments of MAIS DNA are stable in plasmid cloning vectors (e.g. pUC) in *E. coli*.

Table 5.2   Plasmid Content of Selected Strains.

	pLR1/pLR7	*pLR2*	*pLR20*	*large*
LR25	+	+		+
LR163				
LR113	+			
LR540	+		+	+
LR541	+		+	+
LR542	+		+	

3.2   pLR1

The small (17.2 kb) plasmid of strain LR25 (Crawford *et al.*, 1981b) was determined to be related to pLR7 by hybridization. This plasmid has been cloned, as discussed below, and restriction mapped (Figure 5.3; McDermott and Crawford, unpublished). The orientation of the pLR1 map relative to the pLR7 map was confirmed by hybridization of specific pLR7 clones with digests of the pLR1 clones. pLR1 is slightly larger than pLR7 and there is considerable divergence of restriction sites. The HindIII and BamH1 sites of pLR7 are retained, but a second BamH1 site is present in pLR1. The single XbaI site of pLR7 is absent and a single BglII site is present.

3.3   pLR2

We have attempted to clone and restriction map pLR2, the 28 kb plasmid carried by strain LR25. Preliminary restriction analysis demonstrated that this plasmid has two sites each for EcoR1 and BamH1, with each enzyme cutting the plasmid roughly in half. Plasmid DNA from LR25 (a mixture of pLR1, pLR2, and the large plasmid pLR3 — Table 5.2) was prepared by CsCl-ethidium bromide gradient centrifugation, and a library of BamH1 fragments was prepared in lambda EMBL3. pLR2 DNA was excised from an agarose gel, labelled, and used as a probe to identify clones corresponding to pLR2. A total of 50 clones were picked and characterized. All carried the

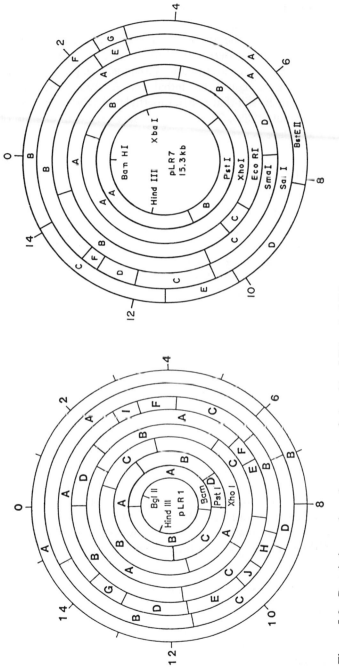

Figure 5.3 Restriction endonuclease maps of plasmids pLR7 and pLR1.

same BamH1 fragment of pLR2. No clones of the other BamH1 fragment were obtained. Since the fragments are of similar size and are produced in equimolar amounts, the difficulty in cloning one of them suggests it may encode some product that is detrimental to *E. coli*, possibly a restriction or modification enzyme as discussed below. Obviously, expression of a cloned restriction endonuclease could be lethal for *E. coli*. The presence of restriction systems in *E. coli* that restrict methylated DNA has recently been reviewed (Blumenthal, 1989). These systems interfere with cloning of some modification methylase genes.

3.4   pLR20, pVT2

Analysis of AIDS-associated strains (Crawford and Bates, 1986) revealed a second small plasmid unrelated to pLR7. A representative of this group, pLR20, is carried in *M. avium* strain LR541 (serotype 4, Table 5.2). This plasmid has not been restriction mapped, but the entire plasmid has been cloned into pUC18 as an EcoR1 fragment. This clone was designated pJC70 (Crawford and Hellyer, unpublished). A closely related plasmid, pVT2, has been cloned and a restriction map developed. pVT2 has unique EcoR1, HindIII, and Pst1 sites. Cloned EcoR1–HindIII fragments have been pooled and used as a hybridization probe to determine if plasmids in clinical and environmental isolates are related. The study demonstrated that plasmid pVT2 hybridized to plasmids from 12.8–15.3 kb in both clinical and environmental isolates (Jucker and Falkinham, in preparation).

3.5   Large plasmids

To date there have been no reports of characterization of the large plasmids other than by size. Although many strains carry similar-sized large plasmids, there is no proof that these are in fact the same or even related. Work with large plasmids has been inhibited by the modest yields that are obtained in most preparations. We were able to isolate only one clone containing a fragment of pLR3 from our lambda library of fragments of strain LR25 plasmid DNA. This probably reflects the low copy number of the large plasmids and their selective loss in the preparative procedure.

3.6   Relationship of the above plasmids

Hybridization analysis indicated that there is little or no sequence homology between pLR7 and related plasmids and the pLR20–pVT2 group of plasmids (Jucker and Falkinham, in preparation). pVT2 and pLR20 (actually the clone pJC70) hybridize with the larger plasmid pLR2. However, there is little or no homology between pLR20 (pJC70) and the cloned BamH1 fragment of pLR2 indicating that the homologous region is in the other (as yet uncloned) half of the plasmid. Analysis of clone pYUB12 (Snapper *et al.*, 1988), which carries the well-characterized *M. fortuitum* plasmid pAL5000 (Labidi *et al.*, 1985), demonstrated that it does not hybridize with pLR7, pLR1, pLR2, or pLR20.

4   DISTRIBUTION OF PLASMIDS IN MAIS STRAINS OF DIFFERENT ORIGIN

4.1   Unselected strains

In a survey of MAIS strains from various sources we demonstrated plasmids in 37 of 130 strains (Crawford *et al.*, 1982). Some of these strains are listed in Table 5.1 and shown in Figure 5.1. Seventy strains were from culture collections and the other 60 were recent isolates from humans and animals. Most of the plasmids were in the 15–30 kb range, but 10 of the strains carried large plasmids. Of the 70 serotyped stock strains 13 were considered *M. scrofulaceum*. Ten of the 13 carried plasmids. Of interest, four of these strains carried large plasmids. Two of these strains, TMC 1307 and 1324, were not serotypable, and a third, TMC 1306, is serotype 9, a serotype normally associated with *M. intracellulare*. Only 12 of the 57 strains of serotypes 1–27 carried plasmids, and there was no obvious correlation between the presence of plasmids and serotype. Most of these strains have been maintained in the laboratory for many years and have maintained plasmids presumably without strong selective pressure. Plasmids were detected in 15 of 60 recent clinical isolates and animal isolates. Multiple plasmids were detected in some of these. We were unable to correlate the presence of plasmids with the source of the strains, serotype, or phage type.

4.2   Environmental strains

In an attempt to identify whether the environment was the source of MAC organisms infecting man, the frequency of plasmids was measured in isolates recovered from human patients (non-AIDS) and water, soil, dust, and aerosol isolates recovered at the same time and within the region of the residence of the patients (Meissner and Falkinham, 1986). The data showed that plasmids were common in clinical isolates (64 of 116; 55%) and aerosol isolates (12 of 16; 75%). Plasmids were less frequently found in isolates recovered from water (10 of 48; 21%), soil (3 of 56; 5%) or dust (0 of 11; <9%). Further, plasmids of environmental isolates have the same molecular weights as those from clinical strains. Although similar-sized plasmids were present in many strains, very few strains had the same overall plasmid profile. Thus, the plasmids provide a very useful epidemiological marker. The results suggest that aerosols generated over waters in the southeastern United States are a likely source of human infection with MAC.

4.3   Clinical strains

Persons with acquired immunodeficiency syndrome are frequently infected with *M. avium* and develop disseminated disease. In the United States the majority of *M. avium* isolates from AIDS patients are of serotypes 1, 4, and 8 (Green *et al.*, 1982; Kiehn *et al.*, 1985). The plasmid content of 26 AIDS-associated isolates was determined (Crawford and Bates, 1986). The strains were of serotypes 4, 8, 4/6, and 4/8. All 26 strains were found to carry plasmids. Southern blot analysis of the intact plasmids probed with labelled cloned fragments of pLR7 revealed that one small plasmid in each of the strains is related to pLR7. The size of the plasmids varied, but all were larger than pLR7. Probing of SalI restriction digests confirmed that the plasmids were related (all fragments hybridized strongly), although considerable divergence in the SalI sites was observed. Fifteen of the strains carry a second small plasmid not related to pLR7. This plasmid was present in most of the serotype 4 strains but in only one serotype 8 strain. As discussed above, representatives of this second small plasmid, pLR20 and pVT2, have been cloned. Ten of the strains carry large plasmids, and these were also most common in serotype 4 or mixed serotype 4/6 and 4/8 strains. In subsequent studies of serotype 1 and untyped strains (unpublished) we have found that most, but not all, AIDS-associated isolates carry plasmids related to pLR7.

Franzblau *et al.*, (1986) analysed 31 MAC clinical isolates from Japan. Plasmids were demonstrated in 16 of these strains. These 16 strains carried from 1–3 plasmids, and six different patterns were noted. Eight strains carried only one large plasmid. Six strains carried small and large plasmids, with two of these strains also having a 52 Mdal (80 kb) plasmid. One strain carried a small plasmid and a 60 Mdal (92 kb) plasmid, and one strain carried only a 44 Mdal (68 kb) plasmid. Hybridization analysis of a few of these strains showed that they carry small plasmids related to pLR7 (Crawford, unpublished).

Similar analysis has been performed on over 100 strains isolated in England (T. Hellyer, personal communication). About one-half of the strains were isolated from AIDS patients and the remainder were other human isolates and veterinary strains. Plasmids were detected in slightly less than half of the strains. Of those carrying small plasmids many hybridized with pJC20 (pLR7) and a small number (mostly serotype 4) hybridized with pJC70 (pLR20). A few strains, not AIDS-associated, carried small plasmids that did not hybridize with either probe. As was seen with the strains from Japan (Franzblau *et al.*, 1986) a number of strains carried only a single large plasmid.

4.4 Animal-derived strains

Recently, Masaki *et al.* (1989) examined the frequency and types of plasmids isolated from MAC strains recovered from lymph nodes of healthy swine. Of 20 strains isolated only four (20%) harboured plasmids suggesting that the porcine lymph nodes contained *M. avium* strains of soil origin. Three of the four strains had two plasmids, one within the common lower size range and one with the higher as discussed above. Interestingly, one plasmid was 4.5 Mdal (6.9 kb) suggesting that some Japanese MAC strains have plasmids smaller than those observed in the United States.

5 PLASMID FUNCTIONS

5.1 Approaches to identifying plasmid-associated functions

Preliminary identification of plasmid functions is usually based upon the comparison of the plasmid content and specific phenotypes in a large number of strains, using known plasmid-encoded phenotypes of other bacteria as a guide. With MAIS strains attempts to associate plasmid DNA content (size and number) with certain characteristics (e.g. antibiotic resistance) have not been very successful because strains harbour multiple plasmids and there is a great deal of phenotypic variation due to differences in chromosomal DNA content in this heterogeneous group. The latter was apparent in comparisons of heavy metal susceptibility between plasmid-free isolates of environmental origin (Falkinham *et al.*, 1984). Definitive identification of plasmid-encoded functions requires the construction of pairs of isogenic strains differing only in their plasmid content. This can be accomplished in one of two ways — elimination of a specific plasmid from a plasmid-carrying strain (curing) or introduction of a specific plasmid into a new host strain. If a plasmid is unstable in its host or if an effective curing agent is available, it is possible to obtain plasmid-free derivatives by brute-force screening of colonies looking for loss of the plasmid without regard to loss of any particular phenotype. However, the procedure is facilitated by the identification of possible plasmid-encoded functions that can be used to screen colonies for loss of the plasmid. The introduction of plasmid DNA into bacteria is generally quite inefficient and identification of a selectable marker is almost always required for transfer of a plasmid into a new host.

The plasmids in MAIS strains are quite stable and curing has proven to be very difficult. We have made numerous attempts to cure strains LR25 and LR113 using various curing agents and have succeeded only once. Agents tested included neutral acriflavin, acridine orange, ethidium bromide, and novobiocin. The general approach has been to prepare a series of broth cultures containing various concentrations of the test agent. After incubation of the cultures for 2–7 days, the culture containing the highest concentration of the agent which is not inhibited (or only slightly inhibited) is plated. Initially, individual colonies were cultured and tested for plasmid content. This approach is obviously limited to screening of modest numbers of colonies. To screen larger numbers we have performed colony lifts and hybridized with cloned plasmid probes (Mazurek et al., 1987). Since the cured derivatives represent the negative (non-hybridizing) colonies among many positive ones only about 200 colonies/plate can be screened. Neither of these approaches has yielded cured derivatives. Recent discovery of conditions which inhibits growth of strain LR25 but not its plasmid-free derivative LR163, may permit identification of conditions for M. avium curing.

Cured derivatives have been isolated following exposure to neutral acriflavin (Crawford et al., 1981b) or as a result of spontaneous loss (Meissner and Falkinham, 1984). Unfortunately, these two pairs of isogenic strains, differing only in plasmid DNA composition, are the only available strains which permit unambiguous assignment of a particular characteristic to a plasmid-encoded function.

5.2   Curing of strain LR25 and plasmid-encoded restriction-modification

As discussed above, M. avium LR25 is a serotype 6 strain that was isolated from a patient with pulmonary disease. Phage typing of this strain suggested the presence of an R-M system (Crawford et al., 1981b). When phage JF2 was propagated on M. smegmatis 607 and then plated onto strain LR25 it gave an efficiency of plating of 10^{-4}–10^{-5} compared with plating back on M. smegmatis. When the rare plaques on strain LR25 were picked and propagated on strain LR25 to obtain high titre stocks they would plate with approximately equal efficiency on both M. smegmatis and strain LR25. Propagation of the phage on M. smegmatis yielded phage that again plated with low efficiency on strain LR25. This is the classical demonstration of restriction and host-induced modification. Since R-M systems are sometimes plasmid encoded, we attempted to isolate a cured derivative using this phenotype as an indicator. A culture of strain LR25 was treated with neutral acriflavin and then plated. Individual colonies were picked and tested for sensitivity to phage JF2. One phage-sensitive clone, designated strain LR163 was identified. Phage propagated in strain 163 is restricted in strain LR25 indicating that both the restriction and modification functions have been lost. Analysis of this strain revealed that it had lost all three plasmids.

To demonstrate that the above observations are due to the presence of an R-M system, lysates of LR25 were prepared and assayed for restriction endonuclease activity using various test DNAs. No activity was detected using pBR322, but the extract cleaved lambda DNA at a single site. The site

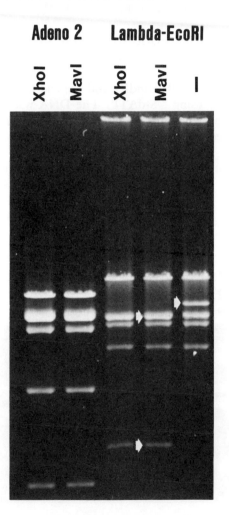

Figure 5.4 Demonstration of restriction endonuclease Mav1. Left two lanes show Adeno 2 DNA digested with Xho1 and Mav1. Right three lanes show Lambda DNA digested with EcoR1 + Xho1, EcoR1 + Mav1, and EcoR1 only.

was located relative to EcoR1 sites and found to correspond to the site for Xhol (Figure 5.4). Digestion of Adeno 2 virus DNA confirmed that the LR25 enzyme is an isoschizomer of Xhol. This enzyme was designated Mav1.

To determine the precise cleavage site we constructed a derivative of pBR322 having a single Xhol site. This was constructed by inserting an Xhol linker into pHP34, an existing derivative of pBR322 having a Smal linker inserted into the EcoR1 site of pBR322 (Prentki and Kirsch, 1982). The structure of the plasmid in the EcoR1 to HindIII region is as follows:

--GAATTCCCCC/TCGAGGGGGAATTC---------GATA/AGCTT---
　　EcoR1　　　　Xhol　　　　　EcoR1　　　　　　HindIII

Plasmid DNA was cut with HindIII and the end filled with large fragment DNA polymerase 1 using ^{35}S-dATP. The DNA was then cut with either Mav1 or Xhol and run on a denaturing 12% acrylamide gel. The same labelled oligonucleotide was produced by both enzymes indicating an identical cleavage site.

Five other *M. avium* strains of known plasmid content were assayed for Mav1 (Figure 5.5, Table 5.1). Plasmid pPM5 which has two Xhol sites was

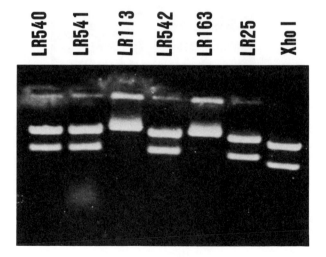

Figure 5.5　Assay of cells extracts for Mav1 activity. Lysates of the indicated strains were assayed for endonuclease activity using pPM5 DNA. This plasmid has two sites for Xhol.

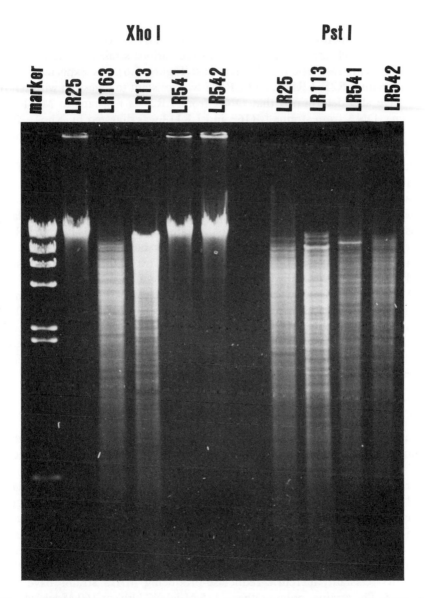

Figure 5.6 Demonstration of specific modification of chromosomal DNA in various MAIS strains. Total chromosomal DNA was digested with Xho1 or Pst1 and electrophoresed on an agarose gel. The marker is a HindIII digest of Lambda.

used for the assay. Mav1 activity was present in extract of strains LR540, 541, and 542. No activity was detected in strain LR163, the cured derivative of strain LR25, or in strain LR113.

The presence of a restriction endonuclease requires a corresponding modification methylase to protect the host DNA. To demonstrate this modification, chromosomal DNA from the various strains was isolated and digested with restriction endonucleases (Figure 5.6). As expected, DNA from strains LR25, LR541, and LR542 was resistant to digestion with Xho1. We have also determined that DNA from LR25 is resistant to Mav1 (not shown). DNA from strains LR163 and LR113 was digested by Xho1. All of the DNAs were sensitive to Pst1 (Figure 5.6), EcoR1, HindIII, and BamH1 (not shown).

It is still not clear which plasmid encodes the Mav1 system. All of the strains except LR163 carry a small plasmid homologous to pLR7, but this plasmid apparently does not encode the R-M system since Mav1 was not detected in strain LR113. Strains LR25, LR540, and LR541 carry large plasmids but strain LR542 does not. Strains LR540, LR541, and LR542 carry pLR20. This plasmid is at least partially homologous with pLR2 of strain LR25 and is the most likely candidate. This would also be consistent with our inability to clone half of pLR2.

To complete the study we isolated DNA from phage JF2 propagated in *M. smegmatis* and tested it for sensitivity to restriction. Surprisingly, the DNA was not cleaved by either Xho1 or Mav1 but was digested by Pst1. In addition, strains LR541 and LR542 are sensitive to phage JF2 but have the Mav1 R-M system. We conclude that a second R-M system must be present in strain LR25 to account for the restriction of phage JF2.

5.3 Plasmid-encoded mercury resistance in *M. scrofulaceum*

A survey of MAIS strains for heavy-metal susceptibility revealed a bimodal distribution of susceptibility to mercury, demonstrating the presence of resistant and susceptible sub-populations (Falkinham *et al.*, 1984). A mercury-resistant, plasmid-carrying strain of *M. scrofulaceum* was isolated and found to volatilize mercury from solution by virtue of mercuric reductase activity. A mercury-sensitive derivative was recovered after growth under non-selective conditions and was found to lack both mercuric reductase activity and a 115 Mdal (177 kb) plasmid present in the parent strain (one of four separate plasmids in this strain; Meissner and Falkinham, 1984). Other mercury-resistant, mercuric reductase-producing MAIS strains have been isolated. The strains contain either a 115 Mdal (177 kb) or 160 Mdal (240 kb) plasmid or lack plasmids entirely (Meissner, 1984), suggesting that the mercury resistance gene may be carried on a transposable genetic element.

Volatilization of mercury leads to enhanced survival of not only myco-bacteria, but other micro-organisms, in mercury-contaminated environments.

5.4 Plasmid-encoded copper resistance in *M. scrofulaceum*

In addition to loss of mercury resistance, the strain of *M. scrofulaceum* lacking the 115 Mdal plasmid was also more susceptible to copper than its parent (Erardi *et al.*, 1987). The copper-resistant parent formed a black, cell-associated precipitate when grown in the presence of copper. The precipitate appeared to be copper sulphide, whose formation is dependent upon the availability of sulphate. Growth of the resistant strain in copper-containing medium resulted in almost complete removal (90%) of copper salts added due to precipitation. Resistance is due to the fact that the copper taken up is precipitated as the sulphide to maintain low intracellular concentrations. Removal and precipitation of copper by this plasmid-associated function leads to mycobacterial survival and detoxification of the environment.

5.5 Role of plasmids in colonial variation

Mizuguchi *et al.* (1981) reported that an opaque colonial variant of *M. avium* had lost a 2.2 Mdal (3.4 kb) plasmid present in the original transparent colony type parent. Though reversible colonial variants of MAIS strains are well known, the authors did not demonstrate if the colonial variation in this particular strain was reversible. Unfortunately, the strain with its unusually small plasmid was lost (Mizuguchi, personal communication).

5.6 Plasmid-influenced catalase activity in *M. avium*

Human clinical and environmental aerosol isolates are able to grow at 43°C (Fry *et al.*, 1986). Further, the ability to grow at 43°C was correlated with the presence of plasmid DNA in all MAC isolates. Using *M. avium* strains LR25 and LR163 (Crawford *et al.*, 1981b), Pethel and Falkinham (1989) demonstrated that the ability of strain LR25 to grow at 43°C was due to resistance to high levels of oxygenation in liquid medium. Because of the probable role of hydrogen peroxide in oxygen-dependent killing, catalase activity was measured on these strains. It was found to be three-times higher in strain LR25 cells in log phase, compared to the activity of the cells of the cured derivative. In addition, cells of LR25 were more resistant to killing by hydrogen peroxide. Both strains produced the same bands of catalase activity on polyacrylamide gels suggesting that they differed in their regulation of catalase activity. Not only did these results suggest a role of plasmid-encoded genes in the regulation of catalase, but they demonstrated the necessity to compare isogenic strains grown to the same stage of growth.

5.7 Role of plasmids in virulence

Establishment of the fact that a high percentage of clinical isolates of MAC strains harboured plasmids (Meissner and Falkinham, 1986), suggested that genes involved in virulence might be plasmid encoded. The observation that most AIDS-associated isolates carry plasmids, especially plasmids related to pLR7, supports that hypothesis (Crawford and Bates, 1986). The striking association of strains of serotypes 1, 4, and 8 with infections in AIDS patients suggests that either these strains are more virulent (or invasive) or there is a greater likelihood of exposure to these strains. The definition of virulence in these strains is difficult, particularly with regard to AIDS. Strains of serotypes 1–3 are virulent in chickens but strains of higher-numbered serotypes show intermediate or no virulence in chickens, rabbits, and mice. Gangadharam *et al.* (1989a) have established the beige mouse model for infection with MAC strains. This animal model shows a number of functional similarities with AIDS in humans, and the mice are highly susceptible to certain MAC strains. The animals can be infected by intravenous, intra-peritoneal, oral, and anal routes. The association of plasmids and virulence was directly tested by assessment of virulence of strain LR25 and strain LR163 in the beige mouse model (Gangadharam *et al.*, 1988). Single cell suspensions of the predominantly transparent colony type were injected intravenously into beige (C57B1/6J-bg^J/bg^J) mice. Virulence was assessed by mortality and changes in the number of organisms in the visceral organs. Strain LR25 caused mortality in 45% of mice challenged and showed marked increases in the number of organisms in the organs. In contrast, no mortality was observed in mice infected with strain LR163 and the number of viable organisms decreased with time. These results were consistent with studies of release of oxygen metabolites from resident and activated macrophages triggered by exposure to the organisms. There is an inverse relationship between virulence and the capacity of *M. avium* to trigger oxygen metabolite release (Gangadharam and Edwards, 1984). Strain LR25 induced the release of low levels of superoxide anion (O_2) and H_2O_2, whereas strain LR163 induced significantly higher levels. The specific functions responsible for these observations are unknown. Specific analysis awaits the preparation of strains differing by only one plasmid or carrying only specific portions of the plasmids.

Gangadharam *et al.* (1989b) have also studied the virulence of the AIDS-associated strains described above (Crawford and Bates, 1986). Although there was not a direct relationship between plasmid content and virulence, there was a correlation between high virulence and the presence of large plasmids. Analysis of this study is not straightforward because of the wide phenotypic variation of MAIS strains and the fact that the strains were unrelated.

Strains of known plasmid content, including the LR25/LR163 pair of strains, were also studied in the human macrophage model (Toba *et al.*, 1989). Only three of 12 strains tested were capable of growth in the macrophages. There was no relationship between plasmid content and growth in macrophages. No differences were seen between strains LR25 and LR163.

6 PROSPECTS

The frequent presence of plasmids in MAIS strains and demonstration of the role of plasmid-encoded genes in functions leading to enhanced environmental survival (i.e. resistance to mercury, copper, and hydrogen peroxide) and pathogenesis demand that increased emphasis be placed upon the study of these genetic elements. The inability to cure strains has limited genetic analysis of these organisms. The development of methods to introduce successfully plasmid DNA into mycobacterial cells (*M. smegmatis* and *M. bovis* BCG) by electroporation (Snapper *et al.*, 1988) offers the opportunity to construct isogenic sets of strains containing representative plasmids. The availability of cloned and mapped small plasmids will make the introduction of selectable markers fairly simple. Hopefully, electroporation can be successfully applied to *M. avium* and will complement attempts to identify methods for curing *M. avium* strains. Transfer of the plasmids will be far superior to curing since a variety of plasmids and modified plasmids could all be studied in a single well-characterized recipient strain. However, transfer of large plasmids may be problematic. We are currently studying the expression in *M. smegmatis* of segments of plasmid pLR7 cloned into vectors derived from plasmid pAL5000 and are attempting to introduce MAC plasmids carrying selectable markers into *M. smegmatis*. Such experiments will allow the identification of individual plasmid gene products and may allow expression of certain phenotypes such as heavy metal resistance or DNA restriction-modification. However, analysis of the role of the plasmids in such complex processes as virulence will require analysis in MAIS strains.

REFERENCES

Birnboim, H. C. and Doly, J. (1979). *Nucleic Acids Res.* 7, 1513–16.
Blasner, M. J. and Cohn, D. L. (1986). *Revs. Infect. Dis.* 8, 21–30.
Blumenthal, R. M. (1989). *Focus* 11, 41–6.
Brooks, R. W., Parker, B. C., Gruft, H. and Falkinham, J. O., III. (1984). *Am. Rev. Respir. Dis.* 130, 630–3.
Crawford, J. T. and Bates, J. H. (1979). *Infect. Immun.* 24, 979–81.

Crawford, J. T. and Bates, J. H. (1984). *Gene* 27, 331–3.

Crawford, J. T. and Bates, J. H. (1986). *Am. Rev. Respir. Dis.* 134, 659–61.

Crawford, J. T., Cave, M. D. and Bates, J. H. (1981a). *Rev. Infect. Dis.* 3, 949–51.

Crawford, J. T., Cave, M. D. and Bates, J. H. (1981b). *J. Gen. Microbiol.* 127, 333–8.

Crawford, J. T., Fitzhugh, J. K. and Bates, J. H. (1981c). *Am. Rev. Respir. Dis.* 124, 559–62.

Crawford, J. T., Deer, P. J., III. and Bates, J. H. (1982). *Ann. Meeting, Am. Soc. Microbiol.*, Abt. C209.

Erardi, F. X., Failla, M. L. and Falkinham, J. O., III. (1987). *Appl. Environ. Microbiol.* 53, 1951–4.

Falkinham, J. O., III, Parker, B. C. and Gruft, H. (1980). *Am. Rev. Respir. Dis.* 121, 931–7.

Falkinham, J. O., III, George, K. L., Parker, B. C. and Gruft, H. (1984). *Antimicrob. Agents Chemother.* 25, 137–9.

Franzblau, S. G., Takeda, T. and Nakamura, M. (1986). *Microbiol. Immunol.* 30, 903–7.

Fry, K. L., Meissner, P. S. and Falkinham, J. O., III. (1986). *Am. Rev. Respir. Dis.* 134, 39–43.

Gangadharam, P. R. J. and Edwards, C. K., III. (1984). *Am. Rev. Respir. Dis.* 130, 834–8.

Gangadharam, P. J. R., Perumal, V. K., Crawford, J. T. and Bates, J. H. (1988). *Am. Rev. Respir. Dis.* 137, 212–14.

Gangadharam, P. R. J., Perumal. V. K., Frhi, D. C. and LaBrecque, J. F. (1989a). *Tubercle* 70 257–71.

Gangadharam, P. R. J., Perumal, V. K., Jairam, B. T., Podapati, N. R., Taylor, R. B. and LaBrecque, J. F. (1989b). *Microbial Pathogenesis* 7, 263–78.

George, K. L., Parker, B. C., Gruft, H. and Falkinham, J. O., III. (1980). *Am. Rev. Respir. Dis.* 122, 89–94.

Good, R. C. (1980). *J. Infect. Dis.* 142, 779–83.

Good, R. C. and Snider, D. E., Jr (1982). *J. Infect. Dis.* 146, 829–33.

Greene, J. B., Sidhu, G. S., Lewin, S., Levine, J. F., Masur, H., Simberkoff, M. S., Nicholas, P., Good, R. C., Zolla-Pazner, S. B., Pollock, A. A., Tapper, M. L. and Holzman, R. S. (1982). *Ann. Intern. Med.* 97, 539–46.

Guerry, P., LeBlanc, D. J. and Falkow, S. (1973). *J. Bacteriol.* 116, 1064–6.

Kado, C. I. and Liu S-T. (1981). *J. Bacteriol.* 145, 1365–73.

Keihn, T. E., Edwards, F. F., Brannon, P., Tsang, A. Y., Maio, M., Gold, J. W., Whimbey, E., Wong, B., McClatchy, J. K. and Armstrong, D. (1985). *J. Clin. Microbiol.* 21, 168–73.

Labidi, A., David, H. L. and Roulland-Dussoix, D. (1985). *Ann. Inst. Pasteur/Microbiol.* 136B, 209–15.

Maniatis, T., Fritsch, E. F. and Sambrook, J. (1982). *Molecular Cloning. A Laboratory Manual.* Cold Spring Harbor Laboratory, NY.

Masaki, S., Konishi, T., Sugimori, G., Okamoto, A., Hayashi, Y. and Kuze, F. (1989). *Microbiol. Immunol.* 33, 429–33.

Mazurek, G. H., Crawford, J. T. and Bates, J. H. (1987). *Ann. Meeting, Am. Soc. Microbiol.*, Abt. U44.

Meissner, P. S. (1984). PhD Dissertation, Virginia Polytechnic Institute and State University, Blacksburg, Virginia.

Meissner, P. S. and Falkinham, J. O., III. (1984). *J. Bacteriol.* 157, 669–72.

Meissner, P. S. and Falkinham, J. O., III. (1986). *J. Infect. Dis.* 153, 325–31.

Mizuguchi, Y. and Tokunaga, T. (1970). *J. Bacteriol.* 104, 1020–1.

Mizuguchi, Y., Fukanaga, M. and Taniguchi, H. (1981). *J. Bacteriol.* 146, 656–9.

Pethel, M. L. and Falkinham, J. O., III. (1989). *Infect. Immun.* 57, 1714–18.

Prentki, P. and Kirsch, H. M. (1982). *Gene* 17, 189–96.

Snapper, S. B., Lugosi, L., Jekkel, A., Melton, R. E., Kieser, T., Bloom, B. R. and Jacobs, W. R., Jr (1988). *Proc. Nat. Acad. Sci. USA* **85**, 6987–91.

Stormer, R. S. and Falkinham, J. O., III. (1989). *J. Clin. Microbiol.* **27**, 2459–65.

Toba, H. Crawford, J. T. and Ellner, J. J. (1989). *Infect. Immun.* **57**, 239–44.

Wendt, S. L., George, K. L., Parker, B. C., Gruft, H. and Falkinham, J. O., III. (1980). *Am. Rev. Respir. Dis.* **122**, 259–63.

6 Plasmids, antibiotic resistance, and mobile genetic elements in mycobacteria

Carlos Martin, Monica Ranes and Brigitte Gicquel

Unité de Génie Microbiologique, Département des Biotechnologies, URA 209 du CNRS, Institut Pasteur, 28 rue du Docteur Roux, 75015, Paris, France

INTRODUCTION

The extensive use of antibiotics for the treatment of bacterial diseases has led to the appearance of resistant strains. The molecular basis of drug-resistance has been studied extensively. In some cases the genetic basis of resistance can be traced to spontaneous chromosomal mutations which result in a modification of bacterial components such as ribosomal proteins, membrane components or enzymes involved in replication, or in transcription. However, in most clinical situations, epidemically-spread antibiotic resistance has been attributed to R plasmids encoding antibiotic inactivating enzymes (e.g. resistance to aminoglycosides), membrane components allowing exclusion of the drugs (e.g. resistance to tetracycline), biosynthetic enzymes resistant to the drugs (e.g. resistance to sulphonamide) or altered target sites (e.g. macrolides). These antibiotic resistance genes have dominant phenotypes and are often located on transposons which are largely responsible for their dissemination.

A summary of current knowledge on the molecular biology of plasmids, antibiotic resistance genes, and mobile genetic elements identified in mycobacteria will be presented in this review. Mycobacteria (of the order actinomycetales) have long eluded genetic analysis, and antibiotic resistance has been presumed to be due to spontaneous mutation or to diminished permeability. The slow growth rates and difficult culturing requirements of mycobacteria, particularly *M. tuberculosis* and *M. leprae*, have hindered progress in genetic studies of these organisms. However tools for gene transfer studies in mycobacteria are now being developed and should provide effective means to study the basis of drug resistance amongst disease-causing mycobacteria. Antibiotic modifying activities have been detected in some mycobacteria, and two antibiotic resistance genes have been isolated and characterized. Genetic studies of mycobacterial plasmids, antibiotic resistance genes, and transposons will be important not only for the rational design of better antibiotic treatment therapies and diagnostic techniques, but for general studies on mechanisms of mycobacterial gene expression and virulence.

It has been proposed that antibiotic resistance genes present in antibiotic-producing streptomycetes (also of the order actinomycetales) may be the origin of resistance genes residing in clinical isolates of bacteria (Benveniste and Davies, 1973; Martin *et al.*, 1988). It is possible that mycobacteria provide a link in the transfer between antibiotic-producing strains and antibiotic-resistant clinical strains.

2 PLASMIDS

Mycobacterial plasmids were first reported in bacteria of the *M. avium complex* and in *M. scrofulaceum* (Meissner and Falkinham, 1986). Some of the plasmids isolated from *M. scrofulaceum* are responsible for resistance to heavy metals, copper, mercury and cadmium (Meissner and Falkinham III, 1984; Erradi *et al.*, 1987, 1989). Resistance to mercuric chloride is associated with a 115 megadalton plasmid since a spontaneously cured derivative failed to grow in the presence of $HgCl_2$ and possessed no detectable mercuric reductase activity. Resistance to copper is linked with the presence of a 173 kilobase (kb) plasmid. Strains harbouring this plasmid remove copper from culture medium as a sulphate precipitate. This phenomenon is not observed in bacterial derivatives lacking the plasmid. In addition, the plasmid bearing strain has a sulphate-independent copper resistance mechanism. Plasmids from the *M. avium* complex were shown to harbour restriction modification functions (see Crawford and Falkinham, this volume).

Plasmids have also been identified in fast-growing mycobacteria, *M. fortuitum* and *M. chelonae*, isolated from patients or from the environment (Labidi *et al.*, 1984). Six plasmids of different molecular weight were found. A 32 kb plasmid was found in most strains. A 112 kb plasmid was shown to be present only in *M. fortuitum var peregrinum*. No biochemical properties or phenotypes could be attributed to any of these plasmids.

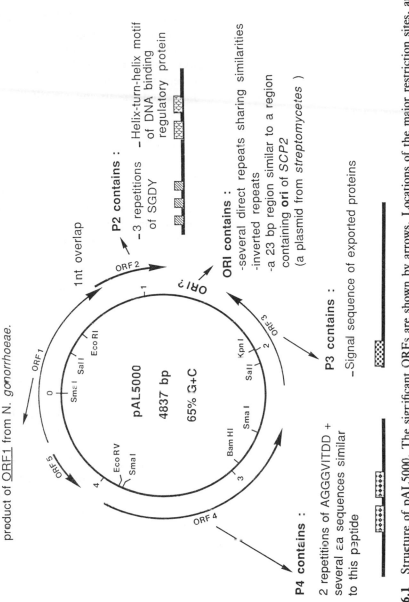

P1 contains : similarities with the translation product of ORF1 from N. gonorrhoeae.

1nt overlap

P2 contains :
- 3 repetitions of SGDY
- Helix-turn-helix motif of DNA binding regulatory protein

ORI contains :
- several direct repeats sharing similarities
- inverted repeats
- a 23 bp region similar to a region containing **ori** of SCP2 (a plasmid from *streptomycetes*)

P3 contains :
- Signal sequence of exported proteins

P4 contains :

2 repetitions of AGGGVITDD + several aa sequences similar to this peptide

ORF 1
ORF 2
ORF 3
ORF 4
ORF 5

ORI?

Sma I
Sal I
Eco RI
Kpn I
Sal I
Sma I
Bam HI
Eco RV
Sma I

pAL5000
4837 bp
65% G+C

0
1
2
3
4

Figure 6.1 Structure of pAL5000. The significant ORFs are shown by arrows. Locations of the major restriction sites, and characteristics of the translation products of the different ORFs are indicated. (Gicquel *et al.*, 1989.)

The most extensively studied mycobacterial plasmid is pAL5000 (5 kb) detected in two *M. fortuitum* strains. The complete nucleotide sequence (4837bp) of this plasmid was determined (Rauzier *et al.*, 1988). It has a 65% G+C content and five open reading frames (orf1 to 5). The general structure of pAL5000 is shown in Figure 6.1. A putative origin of replication was assigned to a 459bp region that has a lower G+C content and that possesses no significant open reading frames. This region contains several features of plasmid origins of replication, direct and inverted repeats and a motif similar to that found near the origin of replication of the streptomycete plasmid SCP2.

Orf1 and orf2 which code for proteins of 389 and 119 aa overlap by one nucleotide and may thus constitute an operon. The translation product of orf1 shows sequence similarities with an orf of pJD1, a plasmid from *N. gonorrhoeae*. The translation product of orf2 contains the characteristic helix-turn-helix motif of DNA binding proteins, and the proximity of this orf to the putative origin of replication suggests that it may play a role in replication. Orf3 may code for an exported protein since the N terminus of its translation product possesses a typical signal sequence. The orf4 translation product contains several repetitions of the consensus motif DDTIVGGGA and shows similarities with the CS surface antigen of plasmodia (for references see Rauzier *et al.*, 1988). According to its location on the physical map this product might correspond to a polypeptide previously identified in *E. coli* minicells containing a pBR322-pAL5000 hybrid plasmid (Labidi *et al.*, 1985).

A functional analysis of this plasmid was performed by gene disruption and deletion experiments. The *E. coli* pTZ19R plasmid and the kanamycin resistant gene from Tn*903* were introduced in the unique restriction sites, KpnI or BamHI, disrupting orf3 and orf4 respectively. The resulting plasmids were able to replicate in *E. coli* and in mycobacteria, *M. smegmatis* and *M. bovis* BCG, showing that both orf3 and orf4 are dispensable for plasmid replication. A deletion removing orf4, orf5 and the N-terminal regions of orf3 and orf1 abolishes plasmid maintenance in mycobacteria, suggesting that orf1 and its 5' region are essential regions of the plasmid. From these results a reduced shuttle vector was constructed that includes a pAL5000 fragment (nt1625 to nt3874), the kanamycin resistance gene from Tn903 and a pUC18 fragment. The different pAL5000 derived shuttle plasmids are shown in Figure 6.2. They have been introduced in *M. smegmatis* and *M. bovis* BCG by electroporation and shown to replicate in these spenies without major rearrangements (Rauzier *et al.*, 1988; Snapper *et al.*, 1988; Gicquel *et al.*, 1989; Ranes *et al.*, *1990*). In mycobacteria, expression of the Tn*903* kanamycin resistance gene seems to be directed from its own transcription and translation start regions (Ranes *et al.*, 1990).

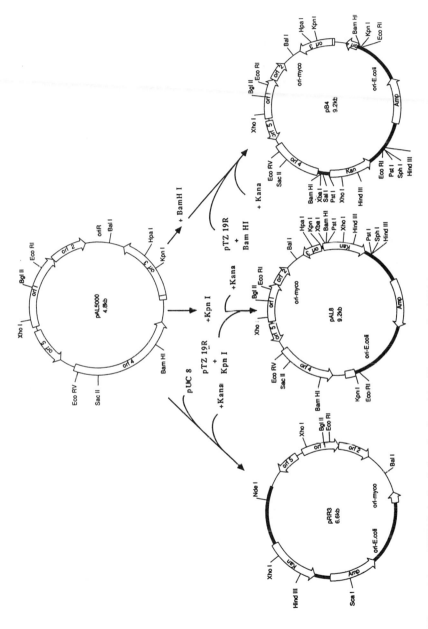

Figure 6.2 pAI5000 derived shuttle plasmids. (Ranes *et al.*, 1990.)

Mycobacterial genes such as the genes encoding the 65 kD protein or OTCase, a protein implicated in the biosynthetic pathway of arginine were shown to be expressed in *E. coli* (Shinnick, 1987; Timm, personal communication). These findings should facilitate the construction of shuttle expression vectors that could be used in *E. coli* and mycobacteria.

3 ANTIBIOTIC RESISTANCE

Sulphonamide was the first antimicrobial drug used for the treatment of bacterial diseases, being introduced for the treatment of tuberculosis in 1935. It was replaced by streptomycin in 1949 and isoniazid was introduced in 1952. Thereafter other drugs were used, rifampicin, *p*-aminosalicylic acid, ethambutol, pyrazinamide, ethionamide and D-cycloserine (McClathy, 1980; Winder, 1982). Despite the increasing number of *M.leprae* strains resistant to dapsone, a sulphonamide derivative, this drug is still used in the treatment of leprosy. Sulphonamides act as analogues of para-amino benzoic acid (PABA), in the synthesis of folate derivatives. Genes encoding dihydropteroate synthase resistant to sulphonamide are present in several transposons described in Gram-negative bacteria (Sundström *et al.*, 1988). However, a mutation affecting the chromosomal gene has been also described in *Streptococcus pneumoniae* (Lopez *et al.*, 1987). Polychemotherapy with dapsone, rifampicin and clofazimine is used as a treatment for leprosy. *M. leprae* strains resistant to rifampicin have been described; mutations in the beta-subunit of RNA polymerase are thought to be responsible. Very few cases of clofazimine resistance have been reported.

At first, streptomycin was very effective in the treatment of tuberculosis. However use of this antibiotic led to the appearance of resistant strains for which mutations in ribosomal protein genes were believed to confer the resistance phenotype. Recently a gene similar to *aphD*, a *Streptomyces griseus* gene encoding an APH6 responsible for resistance to streptomycin, was identified in a *M. fortuitum* strain (unpublished). Despite sequence similarities between these two genes and also with other genes encoding APH6 enzymes, a streptomycin-resistant phenotype was not observed after cloning the mycobacterial APH6 gene in *E. coli* or in *Streptomyces lividans* (unpublished).

M. fortuitum strains are resistant to most antibiotics. In all *M. fortuitum* strains tested, an aminoglycoside acetyltransferase activity was found (AAC). Other aminoglycoside acetyltransferases were found in *M. smegmatis*, *M. phlei* and *M. vaccae* and an aminoglycoside phosphotransferase with activity identical to that of APH(3')IV was found in one *M. fortuitum* isolate (Hull *et al.*, 1984; Udou *et al.*, 1986, 1987). In no case did activities correlate with the pattern of resistance (minimal inhibitory concentration for the different antibiotics) or with the presence of plasmids (only two of eight isolates harbour plasmids). It is possible that high-level resistance requires interactions with other molecules, membrane components, or refractile enzymes such as nucleotidyltransferases or enzymes induced at specific phases of growth in addition to the modifying enzyme.

Mycobacteria such as *M. avium* are resistant *in vitro* to all clinically used antibiotics. Impermeability of the bacterial cell wall is thought to be responsible (Rastogi *et al.*, 1981). The presence of beta-lactamases has been described in most mycobacteria and these enzymes are probably responsible for resistance to beta-lactam antibiotics in this genera (Udou *et al.*, 1986).

4   MOBILE GENETIC ELEMENTS AND ANTIBIOTIC RESISTANCE GENES

Mobile genetic elements are DNA sequences capable of independent excision, that integrate at numerous genomic sites in the absence of DNA sequence homology, and independently of the host recombination system. They were first genetically defined in maize by B. McClintock, and much later found in most organisms which have been examined. Mobile genetic elements were identified in bacteria as insertion mutations with polar effects. Detailed reviews of this subject are available in Berg and Howe (1989). Many of these elements have been shown to harbour genes for antibiotic resistance, determinants of pathogenicity, resistance to heavy metals or catabolic enzymes.

The smallest mobile elements are insertion sequences. They range in length from 800 to 2500 basepairs (bp), and are found in Gram-positive and Gram-negative bacteria and in archaebacteria; they are present on chromosomes, plasmids or bacteriophages. For instance, the well-studied IS*1* sequence is present in 6–10 copies in *E. coli*, while Shigella species may carry more than 50 copies. Many IS elements are present in high copies in archaebacteria, and are thought to be responsible for the genetic instability seen in these organisms.

The structure of insertion sequences is simple; they contain only the necessary transposase functions and short inverted repeats at their extremities. After transposition, 2–12 bp of the target DNA are found as direct repeats flanking the IS at its new location. Several IS sequences are present as components of compound transposons containing antibiotic resistance genes flanked by the IS elements in direct or reverse orientation. Either one IS copy or the entire transposon is involved in the mechanism of transposition. Other transposons are composed of antibiotic-resistance genes and transposase functions flanked by short inverted repeats; Tn*3* is the prototype of such

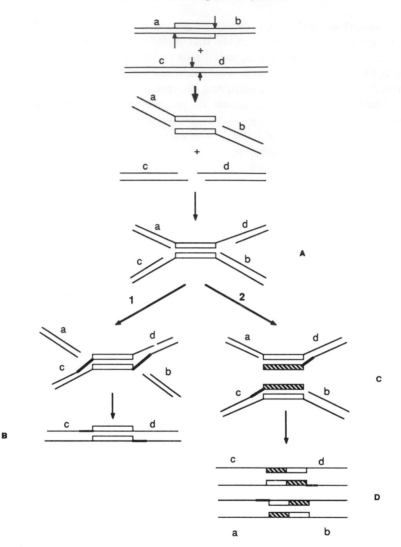

Figure 6.3 Formation and resolution of the Shapiro intermediate. An example of three possible outcomes is illustrated. The transposon, represented by an open rectangle, is cleaved at its 3' extremities. A staggered cleavage of the target site leads to 3' protruding ends. Joining of the 5' ends of the transposon with the 3' ends of the target forms a transposition intermediate (A) which can be resolved in two ways: (1) cleavage of the 5' ends of the transposon and replication of the target site (thick lines) leads to a simple insertion of the transposon at the target site, with a repetition of this site (B); (2) replication of the target site, and of the transposon (dashed rectangles) leads to the formation of a cointegrate (C) that can be resolved either by a site specific resolvase or by the general homologous recombination pathway of the host (D).

elements. Some of these elements do not possess specific sites of insertion. For instance, Tn5 and Tn10 can insert randomly in the chromosome. In contrast, transposons such as Tn7 and Tn554 have a precise specificity for insertion; Tn917 has a hot spot for insertion in *Bacillus subtilis*.

Several models have been proposed to explain transposition. Based on the products obtained after transposition two mechanisms (1) replicative transposition or (2) conservative transposition have been proposed. Lysogenization by Mu bacteriophage occurs via a conservative transposition and its replication by a series of replicative transpositions (for a review see Pato, 1989). *In vitro* studies of the transposition mechanism with purified Mu components have substantiated the model suggested by Shapiro (1979). The intermediate complex of transposition is shown in Figure 6.3. Depending on the resolution of this complex, Mizuuchi (1983) has proposed that transposition could be conservative (Tn10) or replicative (Tn3). Replicative transposition leads to the formation of cointegrates that can be resolved either by a site specific resolvase encoded by the transposon, or by homologous recombination mediated by the host. The different steps are represented in Figure 6.3.

4.1 Tn*610*, a sulphonamide resistance transposon

In the course of cloning a mycobacterial antibiotic resistance gene, Tn610 was isolated from a *M. fortuitum* strain (Martin *et al.*, 1990). Tn610 has 4070bp and is composed of a central fragment homologous to sequences found on transposons of the Tn21 family. Tn610 is flanked by insertion sequences IS6100, that share similarities with the IS6 family (Figure 6.4). Different members of the Tn21 family are aligned in Figure 6.5 They differ from each other by the insertion of different antibiotic resistance genes at an insertional hot spot located between *sul* and *tnpI*. Like others of the Tn21 family, the central fragment of Tn610 contains the sulphonamide resistance gene and the site-specific integrase. However the divergent region corresponding to the different antibiotic resistance genes is missing in Tn610 (Figure 6.4). Sundström *et al.* (1988) have proposed recently that an ancestral transposon of this family would contain only one resistance gene encoding resistance to sulphonamide, the first antimicrobial drug ever used. The insertion of different antibiotic resistance genes between *sul* and *tnpI* might have resulted from the action of the *tnpI* gene product, as a site specific integrase. The centre of Tn610 may correspond to a part of such a transposon.

IS6100 is a 880bp sequence which possesses a large open reading frame of 254aa that shares homologies with the transposases of the IS6 family. Inverted repeats of these different elements also have similarities, suggesting

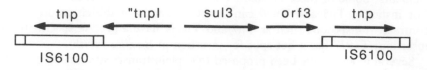

Figure 6.4 General structure of Tn*610*.

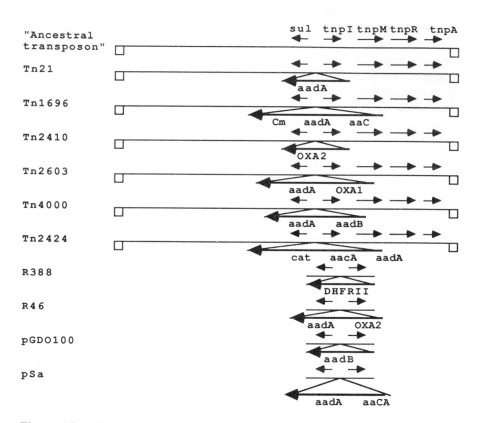

Figure 6.5 Alignment of Tn*21*-like transposons and related plasmids. The different transposons contain the following genes: *sul* (sulphonamide resistance gene), *tnpI* (site-specific integrase), *tnpM*, (modulator of the resolvase), *tnpR* (resolvase) and *tnpA* (transposase). The related plasmids contain *sul* and *tnpI*. As suggested by Sundström *et al.* (1988), all these related structures might have derived from an ancestral transposon (indicated above the Tn*21* family sequences) by insertion of several antibiotic resistance genes between *sul* and *tnpI*.

a common mechanism of transposition. Comparisons of inverted repeats and transposases are shown in Figures 6.6 and 6.7. Regulatory regions of the different transposases show similarities; -35 and -10 regions are conserved except in the mycobacterial sequence where the -10 consensus AATAAT is replaced by GAGCAT, a sequence already encountered for a -10 region of a streptomycete promoter (Ranes, 1988). G+C content and codon usage of the IS6100 orf are characteristic of mycobacterial sequences suggesting that the sulphonamide region has been mobilized from an ancestral transposon by a mycobacterial IS. Four copies of IS6100 are present in the chromosome of *M. fortuitum* strain FC1. This IS was not detected in other mycobacteria tested (*M. bovis* BCG, *M. tuberculosis* or other *M. fortuitum* strains). However transposition with Tn610 has been demonstrated in *M. smegmatis*. Transposition occurs by replicative transposition of one of the insertion sequences and as already observed for others IS members of the IS6 family, it leads to the formation of a cointegrate. A schema of this process is shown in Figure 6.8.

Both ends IS6100	T T T G C A A C A G A G C C
Both ends IS240	T T T G C A C C A G A A C C T T
Both ends IS26	T T T G C A A C G T G C C
Both ends IS51	T A A C T T T G C A A C G A A C C
Right end IS431R C T A T A A A T T C A A C	T T T G C A A C A G A A T C

Figure 6.6 Comparison of nucleotide sequences of the terminal inverted repeats of several IS6 family sequences: IS6100, IS240, IS26, IS51 and IS431. Conserved residues are underlined. References are indicated in Martin *et al.* (1990).

4.2 IS-like sequences

Two mycobacterial sequences sharing similarities with known insertion sequences have been described. However transposition has not yet been demonstrated.

4.2.1 IS900

IS900 was isolated from the clone pMBZ2, derived from a genomic library prepared from a human Crohn's disease isolate of *M. paratuberculosis* (McFadden *et al.*, 1987; Green *et al.*, 1989). It is a 1451bp sequence which is repeated with the same genomic distribution, 15–20 times in all *M. paratuberculosis* strains. It lacks terminal inverted and direct repeats characteristic of most insertion sequences, but shows similarities with IS110, and the

```
IS6100   MTDFKWRHFQGDVILWAVRWYCRYPISYRDIEEMLAERGISVDHTTIYRWVQCYAPEMEKRLRWFW      (66)
IS240 B  MEKE--NIFKWKHYQADMLWTWVRWYLRYNLSFRDIVEMEERGISLSHTTIMRWVHQYGPEINERIRKHL   (69)
IS26     MNPFKGRHFQRDIILWAVRWYCKYGISYRELQEMLAERGVNVWESTIYRWVQRAPEMEKRLRWYW       (66)
IS1      MNYFKGKQFQKDVIIVAVGYLLRYNLSYRELQELLYDRGINVCHTTIYRWVQEMSKVIYHLWKKKN      (66)
IS431    MNYFRYKQFNKDVITVAVGYLRYALSYRDISEIIRERGVNVHHSTVYRWVQEYABIIYQIWKKKH       (66)

IS6100   RRGFDPS--WRLDETYVKVRGKWTYLYRAVDKRGTIDFYLSPTRSAKAAKRFIGKAIIRGLKHWEKPATEN  (135)
IS240 B  KSTNDS---WRVEETYIKIKGENMYLYRAVDSEGNTLDFYLSKKRDEKAAKCFLKKAIASFHVTK-PRVIT  (136)
IS26     RNPSDLCPWHMDETYVKVNGRWAYLYRAVDSRGRTVDFYLSSRRNSKAAYRFIGKIIENNVKKWQIPRFEN  (136)
IS1      RQSFYS---WKMDETYIKIKGRWHYLYRAIDADGLTLDIWLRKKRNTQAAYAFLKRIHKQFGQ---PRVTV  (131)
IS431    KKAYYK---WHIDETYIKIKGKWSYLYRAIDAEGHTLDIWLRKQRDNHSAYAFIKRLIKQFGKPQ--KVIT  (132)

IS6100   TDKAPSYGAATELKREGKIDRETAHRQVKYLNNVIEADHGKLKILIKPVRGFKSIPTAYAIIKGFEVMR    (205)
IS240 B  VDGNKAWPVAIRELKNFKSISYGMPLRVKKYLNMIEQDHRFIKKRIRNMLGLKSMQTAVKMIAGTEAMH    (206)
IS26     TDKAPAYGRAIALLKREGRCPSDVEHRQIKRRNNVIEQHGKLKRIIGATLGFKSMKTAYAIIKGIEVMR    (206)
IS1      TDKAPSLGSAFRKLQSNGLYTKTEHRTV--KTLNNLIEQDHRPIKRPNKYRSLR---TASTTIKGMETIR   (197)
IS431    --DQAPSTKVAMAKVIKAFKIKPDCHCTS--KILNNLIEQDHRHIKVRKTRYQSIN---TAKNTLKGIECIY  (197)

IS6100   ALRKGQARPWCLQPGIRGEVRLVERAFGIGPSALTEAMGMINHHFAAAA    (254)
IS240 B  MVKKGQLKLRAQ--------SAQNQNRCIHQ--FGLTA               (235)
IS26     ALRKGQASAF--------YYGDPLGEMR--IVSRVFEM               (234)
IS1      GIYKKNRRNGTL--------FGFSVSTEIKV-IMGILA               (226)
IS431    TEMKKNRSLQI--------YGFSPCHEISIMAS                    (224)
```

Figure 6.7 Comparison of deduced amino-acid sequences of several transposases of the IS6 family: IS6100, IS240, IS26, IS1 and IS431. They are aligned by using the program of Saurin and Marlière (1986). The conserved amino acid residues are indicated by capital letters. Gaps are indicated by spaces. (Martin et al., 1990.)

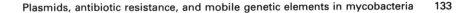

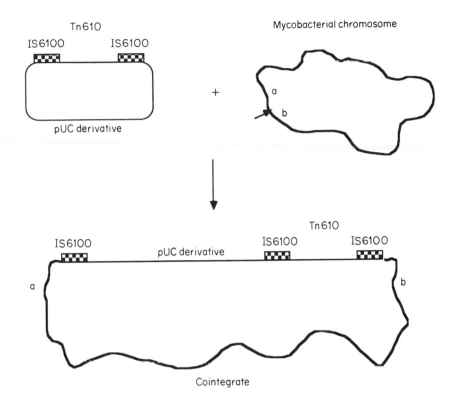

Figure 6.8 Transposition of IS*6100* between markers a and b of a mycobacterial chromosome. Insertion of IS*6100* occurs via a replicative transposition process which leads to the formation of a cointegrate.

minicircle, transposable elements from *Streptomyces coelicolor*. Absence of inverted and direct repeats has also been observed for Tn*554*, a transposon with specific insertion sites, isolated from *Staphylococcus aureus*. A single orf coding for a predicted 399aa polypeptide shares homologies with orf1215 from IS*110* and orf1 from the minicircle (Figure 6.9). An analysis of the sequences flanking IS*900* in three different clones and of one of the unoccupied loci of the closely related strain *M. avium*, has revealed a potential consensus recognition site for IS*900* insertion (Figure 6.10). Sequences related to IS*900* have been detected in *M. avium* strains which, like *M. paratuberculosis* strains, depend upon mycobactin for growth (McFadden *et al.*, submitted). IS*900* DNA probes have been used to characterize *M. paratuberculosis* strains by restriction fragment length polymorphism (RFLP) analyses in Southern blot experiments. IS*900* specific

```
IS900 ORF1197   vaqp...VWAGVDAGKrADHycmvinddaqr.LLSQRva......NDEAALLEliaa.........VTTLADGGevtwAIDLNAGGAALLIALLIAAG
IS110 ORF1215   mfdtedvgVFLGLDVGKTAHaghgltpagkkvLDKQLp......NSEPRLRAvfdklaakfgtvlviVDQPASIGalplTVARDAGCKVAYLPGLAMRR
Minic ORF1      mwedslt.VFCGIDWAERHHdvaivddtgtl.LAKARitddvagyNKLLDLLAehgdssatpip...VAIETSHGliv.AALRTGSRKVFAINPLAAAR

IS900 ORF1197   QRLLYIPGRtvhaagsyrgegKTDAKDAAIIADQARmr.HDLQPLRAGDDIAVELRILtsrrSDLVADRTRAIEPNARPaagils.ALERaf..DYn..
IS110 ORF1215   IADLYPGEA...........KTDAKDAAVIADAAMa.HTLRSLELTDEITAELSVLvgfdQDLAAEATRTSNRIRGLltqfhp.SLERv..LGp..
Minic ORF1      YRDRHGVSRk..........KSDPGDALVLANILRtdmHAHRPLPADSELAQAITVLaraqQDAVWNRQQVANQVRSLlreyypaALHAfqskDGglt

IS900 ORF1197   KSRAALILLTGYq.......TPDALRSAGGA......RVAAFl..RKRKARNADTVAATALQAAnaqhsiVPGQQLaatVVARLAKEVMALD....TEi
IS110 ORF1215   RLQAVTWLLERYg.......SPAALRKAGRRrivelvRPKAP..RMAQRLIDDIFDALDEQTVv.....VPGTGDi..VVPSLASSLTAVHeqrraLE.
Minic ORF1      RPDARVILTMAPtpakaaklTLAQLRAGLKRsg...RTRAFnteIERLRGIFRSEYARQLPAVed...AFGHQLlalLRQLDATCLAADD....LAk

IS900 ORF1197   GDTDAMIEERFRrhrhaeiILSMPGFGVILGAEFLaatGGDMAAFASADRLAGVAGLAPVPRDSGRi.SGnlkrPRRyd..RRLLRACYLSALVsirTDP
IS110 ORF1215   AQINALLEAHPSlpv....LTSMPGVGVRTAAVLl...VTVGTSFPTAAHLASYAGLAPTTKSSGTsiHGeha.PRGgn..RQLKRAMFLSAFAcmnADP
Minic ORF1      AVEDAFREHADSei.....LLSFPGLGPLLGARVLaeiGDDRSRFTDARALKSYAGSAPITRASGRk.HFv...GRRfvknRLMNAGFLWAFAalqASP

IS900 ORF1197   .SSRtyYDRKRTEGKRHTQAVLALARRRLNVLWAMLr...DHAVYHPATTTAAA.
IS110 ORF1215   .ASR..YDRQRARGKTHTQALLRLARQRISVLFAMLrdgt.FYESRMPAGVELAA.
Minic ORF1      gANAh.YRRRREHGDWHAAAQRHLLNRFLGQLHHCLqtrqhFDEQRAFAPLLQAAa
```

Figure 6.9 Comparison of the major IS900 ORF (ORF1197) with ORFs found on transposable sequences from *Streptomyces coelicolor* A3(2): ORF1215 from IS110 (Bruton and Chater, 1987), and ORF1 from the minicircle (Henderson *et al.*, 1989). Sequences are aligned by using the program of Saurin and Marlière (1986). Conserved amino acids are indicated by capital letters. Gaps are indicated by spaces.

oligonucleotides can be used for the detection of *M. paratuberculosis* in veterinary samples in *in vitro* DNA amplification experiments (see McFadden, this volume).

```
ATGGTCATGGTGGtcctt.......IS900.......gagaatCCCCTTGGCA

AACGACATGTGTTtcctt.......IS900.......gagaatCCCCTTACGC

TGGTCATGTGGTGTtcctt.......IS900.......gagaatCTCCTTCGCG
```

Figure 6.10 Comparison of three sites of insertion of IS*900*. Extremities of IS*900* are indicated in lower cases. Target sites are indicated in capital letters; conserved nucleotides are underlined.

4.2.2   *IS*6110

From a cosmid library of *M. tuberculosis*, strain H37rv, IS*6110* was isolated as a repetitive sequence that has similarities to IS*3411*, an IS from the *E. coli* transposon which carries genes for citrate utilization (Thierry *et al.*, 1990). IS*6110* (1340bp) possesses, at its extremities, 28bp inverted repeats homologous to those found in IS*3411* and IS*3*; 3bp direct repeats are found in the flanking DNA. Such repeats might have arisen by repetition of the target sequence during the transposition event. IS*6110* contains two significant orfs with similarities to the orfs of IS*3411*. One of them might encode the transposase since it is similar to the transposases of IS*3411* and IS*3* (unpublished). Southern blot experiments with genomic DNAs hydrolysed with several restriction enzymes have shown that IS*6110* is present at least 15 times in *M. tuberculosis* and *M. microti*, but only one or two times in *M. bovis* and *M. bovis* BCG. Southern blots performed after pulsed field electrophoresis experiments have shown that IS*6110* is scattered throughout the genome of *M. tuberculosis* (unpublished). IS*6110* is present in H37rv, H37ra and in all *M. tuberculosis* strains isolated from patients. Two oligonucleotides derived from the IS*6110* DNA sequence were used to screen the presence of IS*6110* in other mycobacterial strains. IS*6110* was detected in all mycobacteria of the *M. tuberculosis* complex but not in mycobacteria that do not belong to this complex (unpublished). Cross-hybridization was observed between IS*6110* and a repeated sequence isolated from *M. tuberculosis* (Eisenach *et al.*, 1988). Sequence data available from the latter and IS*6110* indicate that they are identical. Probes from IS*6110*-derived sequences can be used for the detection of strains from *M. tuberculosis* complex directly in clinical specimens by *in vitro* DNA amplification (unpublished).

4.3 Other repetitive DNA sequences

Variable repetitive sequences (at least 28 copies) were identified in *M. leprae*: a 545bp central domain with either a 100bp left-end and a 44bp right-end with a 47bp extention has been found in some of these elements (Cole *et al.*, unpublished results). These sequences are present with the same distribution pattern in the four different *M. leprae* isolates that have been examined (Clark-Curtiss and Docherty, 1989; Grosskinsky *et al.*, 1989; see Clark-Curtiss, this volume). One of these sequences is located 3′ to the gene encoding the 65 kD heat shock protein. No significant orfs were identified and no homology with any other known sequences was observed.

5 DETECTION OF ANTIBIOTIC RESISTANCE

Conventional methods for susceptibility testing of mycobacteria involve 3–4 weeks of culturing on solid media containing the various antibiotics to be tested. Detection of *M. leprae* strains resistant to dapsone or rifampicin requires inoculation of mouse foot pads and administration of the antibiotic for at least three months.

Methods which assess drug susceptibility in shorter periods of time (1–2 weeks) have been developed. BACTEC systems involve measurement of $^{14}CO_2$ produced by bacteria growing in broth containing ^{14}C-labelled palmitic acid and antimicrobial agents (Laszlo and Siddiqi, 1984; Kirihara *et al.*, 1985; Rastogi *et al.*, 1989). Tests using bioluminescence for quantification of ATP production also allow the detection of mycobacterial growth in 1–2 weeks (Nilsson *et al.*, 1988).

For identification, methods based on the use of DNA probes have been used successfully on cultures (Roberts *et al.*, 1987; Drake *et al.*, 1987; Kiehn and Edwards, 1987; Kawa *et al.*, 1989). Detection of mycobacteria for diagnostic purposes will be improved by the techniques of *in vitro* DNA amplification. A variety of primers and probes have been developed recently and evaluated for the rapid detection of mycobacteria in clinical specimens (Hance *et al.*, 1989; Brisson-Noël *et al.*, 1989; Thierry *et al.*, 1990 and unpublished results; see McFadden *et al.*, this volume). Isolation and nucleotide sequencing of genes involved in antibiotic resistance will provide additional sequence information useful for rapid identification of antibiotic resistance directly in clinical specimens, without prior isolation and culturing of the mycobacteria.

ACKNOWLEDGEMENTS

We are grateful to J. Davies for critical advice and help with the manuscript. We acknowledge S. Cole and J. McFadden for providing us with unpublished information. This work was supported by the UNDP/World Bank/ WHO special programme for research and training in tropical diseases and Institut Pasteur.

REFERENCES

Benveniste, R. and Davies J. (1973). *Proc. Natl. Acad. Sci., USA* **70**, 2276–80.
Berg, D. E. and Howe, M. M. (1989). *Mobile DNA*, American Society for Microbiology, Washington, DC.
Brisson-Noël, A., Gicquel, B., Lecossier, D., Lévy-Frébault, V., Nassif, X. and Hance, A. J. (1989). *Lancet* **ii**, 1069–71.
Bruton, C. J. and Chater, K. F. (1987). *Nucleic Acids Res.* **17**, 7053–65.
Clark-Curtiss, J. E. and Docherty, M. A. (1989). *J. Infect. Dis.* **159**, 7–15.
Drake, T. A., Hindler, J. A., Berlin, G. W. and Bruckner, D. A. (1987). *J. Clin. Microbiol.* **25**, 1442–5.
Eisenach, K. D., Crawford, J. T. and Bates, J. H. (1988). *J. Clin. Microbiol.* **26**, 2240–5.
Erardi, F. X., Failla, M. L. and Falkinham III, J. O. (1987). *Ap. Env. Microbiol.* **53**, 1951–4.
Erardi, F. X., Failla, M. L. and Falkinham III, J. O. (1989). *Antimicrob. Agents. Chemother.* **33**, 350–5.
Gicquel-Santez, B., Moniz-Pereira, J., Gheorghiu, M. and Rauzier, J. (1989). *Acta Leprologica* **7**, 208–11.
Green, E. P., Tizard, M. L. V., Moss, M. T., Thompson, J., Winterbourne, D. J., McFadden, J. J. and Hermon-Taylor, J. (1989). *Nucleic Acids Res.* **17**, 9063–73.
Grosskinsky, C. M., Jacobs, W. R., Jr, Clark-Curtiss, J. and Bloom, B. R. (1989). *Infection and Immunity* **57**, 1535–41.
Hance, A. J., Grandchamps, B., Lévy-Frébault, V., Lecossier, D., Rauzier, J., Bocart, D. and Gicquel, B. (1989). *Molec. Microbiol.* **3**, 843–9.
Hull, S. I., Wallace, Jr, R. J., Bobey, D. G., Price, K. E., Goodhines, R. A., Swenson, J. A. and Silcox, V. A. (1984). *Am. Rev. Respir. Dis.* **129**, 614–18.
Henderson, D. J., Lydiate, D. J. and Hopwood, D. A. (1989). *Molecular Microbiol.* **3**, 1307–18.
Kawa, D. E., Pennell, D. R., Kubista, L. E. and Schell, R. (1989). *Antimicrob. Agents. Chemother.* **33**, 1000–5.
Kiehn, T. E. and Edwards, F. F. (1987). *J. Clin. Microbiol.* **25**, 1551–2.
Kirihara, J. M., Hillier, S. L. and Coyle, M. B. (1985). *J. Clin. Microbiol.* **22**, 841–5.
Labidi, A., Dauguet, C., Goh, K. S. and David, H. L. (1984). *Cur. Microbiol.* **11**, 235–40.
Labidi, A., David, H. L. and Roulland-Dussoix, D. (1985). *FEMS Microbiol. Lett.* **30**, 221–25.
Laszlo, A. and Siddiqi, S. H. (1984). *J. Clin. Microbiol.* **19**, 694–8.
Lopez, P., Espinosa, M., Greenberg, B. and Sanford, A. L. (1987). *J. Bacteriol.* **169**, 4320–6.
McClathy, J. K. (1980). In *Antibiotics in Laboratory Medicine*, (ed. V. Lorian), pp. 135–69. Williams and Wilkins, Baltimore.
Martin, C., Timm, J., Rauzier, J., Gomez-Lus, R., Davies, J. and Gicquel, B. (1990). *Nature* (in press).
Martin, P., Jullien, E. and Courvalin, P. (1988). *Molec. Microbiol.* **2**, 615–26.

McFadden, J. J., Butcher, P. D., Chiodini, R. and Hermon-Taylor, R. (1987). *J. Clin. Microbiol.* **25**, 796–801.

Meisner, P. S. and Falkinham III, J. O. (1984). *J. Bacteriol.* **157**, 669–72.

Meisner, P. S. and Falkinham III, J. O. (1986). *J. Inf. Dis.* **153**, 325–331.

Mizuuchi, K. (1983). *Cell* **35**, 785–94.

Nilsson, L. E., Hoffner, S. E. and Anséhn, S. (1988). *Anti. Ag. Chem.* **32**, 1208–12.

Pato, M. L. (1989). In *Mobile DNA* (eds D. E. Berg and M. M. Howe), pp. 23–52. American Society for Microbiology, Washington, DC.

Ranes, M. (1988). Ph.D thesis Harvard University, MA.

Ranes, M. G., Rauzier, J., Lagranderie, M., Gheorrghiu, M. and Gicquel, B. (1990). *J. Bacteriol.* **172**, 2793–97.

Rauzier, J., Moniz-Pereira, J., and Gicquel-Sanzey, B. (1988). *Gene* **71**, 315–21.

Rastogi, N., Frehel, C., Ryter, A., Ohayon, H., Lesourd, M. and David, H. L. (1981). *Anti. Ag. Chem.* **20**, 666–77.

Rastogi, N., Goh, K. S. and David, H. L. (1989). *Res. Microbiol.* **140**, 405–17.

Roberts, M. C., McMillan, C. and Coyle, M. B. (1987). *J. Clin. Microbiol.* **25**, 1239–43.

Saurin, W. and Marlière, P. (1986). *C. R. Acad. Sci. Paris* **13**, 541–6.

Shapiro, J. A. (1979). *Proc. Natl. Acad. Sci., USA* **76**, 1933–7.

Shinnick, T. M. (1987) *J. Bacteriol.* **169**, 1080–8.

Snapper, S. B., Lugosi, L., Jekkel, A., Melton R. E., Kieser, T., Bloom, B. R. and Jacobs, R., Jr (1988). *Proc. Natl. Acad. Sci., USA* **85**, 6987–91.

Sundström, L., Radström, P., Swedberg, G. and Sköld, O. (1988). *Mol. Gen Genet.* **213**, 191–201.

Thierry, D., Cave, M. D., Eisenach, K. D., Crawford, J. T., Bates, J. H., Gicquel, B. and Guesdon, J-L. (1990) *Nucleic Acids Res.* **18**, 188.

Udou, T., Mizuguchi, Y. and Yamada, T. (1986). *Am. Rev. Res. Dis.* **133**, 653–7.

Udou, T., Mizuguchi, Y. and Wallace, Jr, R. J. (1987), *Am. Rev. Res. Dis.* **136**, 338–43.

Udou, T., Mizuguchi, Y. and Wallace, Jr, R. J. (1989). *FEMS Microbiol. Lett.* **57**, 227–30.

Winder, F. G. (1982). In *The Biology of the Mycobacteria*, Vol. 1, (ed. C. Ratledge) pp. 354–438. Academic Press, London.

7 DNA probes for detection and identification

Johnjoe McFadden, Zubair Kunze and Patrick Seechurn

Department of Microbiology, University of Surrey, Guildford, Surrey, GU2 5XH, UK

1 PRINCIPAL OF NUCLEIC ACID HYBRIDIZATION

DNA and RNA hybridization follow essentially the same principles and therefore we will simplify the text by referring only to DNA; however, most of what will be discussed will equally well apply to hybridizations involving RNA.

1.1 DNA denaturation

When an aqueous solution of double-stranded DNA is heated, the UV absorbance increases with increasing temperature. This is due to local reversible denaturation, or looping out, of short stretches of DNA, resulting in a reduction in hyperchromicity and increased absorbance. At low temperatures this effect is reversible; the DNA will spontaneously reanneal on cooling and the absorbance will return to the initial value. However if the temperature increases beyond the 'melting temperature', T_m, then the denaturation becomes irreversible and the strands separate. When the solution is cooled, the absorbance remains high. The melting temperature of dsDNA is a function of the thermal stability of DNA, which in turn is a function of the number of base pairing interactions per unit length. In perfectly homologous DNA this will be a function of the base ratio (G+C content) of the DNA, since GC pairing involves three hydrogen bonds, whereas AT pairing involves only two. The T_m is also affected by the ionic

MOLECULAR BIOLOGY OF THE MYCOBACTERIA
ISBN 0–12–483378–0

strength of the medium (being raised with increasing ionic strength, e.g. salt concentration), the presence of ions (e.g. Mg^+), polysaccharide, protein and organic solvents (e.g. formamide). However, under standard conditions, measurement of T_m can be used to estimate G+C content of DNA and has been widely used to estimate base ratios for mycobacterial DNA (see Clark-Curtiss, this volume).

The T_m can also be used to estimate DNA homology in heterologous DNA hybrid (produced by DNA hybridization — below), since in this case the thermal stability will depend on the number of homologous basepairing interactions in the hybrid and therefore the amount of DNA homology between the parent strands.

1.2   DNA reassociation

The reassociation of denatured double-stranded DNA, at temperatures below the T_m, is a spontaneous reaction. The rate-limiting step is the collision of the complementary DNA strands and therefore the reaction follows approximately second-order kinetics with the reaction rate proportional to the product of the concentration of complementary DNA. For homopolynucleotides (e.g. polydG/polydC), this is a measure only of the DNA concentration. However, for complex DNA the concentration of complementary DNA is inversely proportional to the complexity of DNA (single copy genes are at a lower concentration in complex genomes) and therefore the rate of renaturation of DNA decreases as the complexity increases — small genomes renature faster than large genomes. Genome size may therefore be estimated from measurements of rates of DNA renaturation; and mycobacterial genome sizes have usually been estimated using this technique (see Clark-Curtiss, this volume). The measurement of DNA renaturation kinetics may also be used to detect repetitive DNA in complex genomes; since the concentration of repetitive DNA is relatively higher, it will hybridize at a higher rate than single-copy DNA.

1.3   DNA hybridization

Most uses of DNA probes involve strand annealing between heterologous DNA, a reaction known as DNA hybridization. The formation of stable DNA hybrids between heterologous DNA strands will be possible only at temperatures below the T_m of the hybrids. As described above, the T_m of heterologous DNA is largely a function of the number of basepairing interactions and therefore the homology between the strands. Since the T_m is affected by all of the factors mentioned above (ionic strength, organic solvents), DNA hybridization will be similarly affected. Stringency is a loosely defined term encompassing all the factors affecting hybrid stability: temperature, salt concentration, presence of organic solvents. Conditions of low stringency (e.g. high salt, low temperature) allow DNA hybrids to form between DNA species with low levels of DNA homology, whereas high stringency conditions (low salt, high temperature, presence of e.g. formamide) allow hybridization only between DNA species with high levels of homology. This simple principal, coupled with methods for detection of hybrid is the basis of most DNA probe applications.

1.4 Hybrid detection

One of the simplest methods to detect hybrid is to utilize the hyperchromicity of DNA to monitor, spectrophotometrically the conversion of ssDNA to dsDNA (e.g. Baess and Weis Bentzon, 1978; Baess, 1979, 1983, 1984). However, the requirement for large amounts of DNA, a spectrophotometer with thermostatically controlled cuvette chambers and the inability to monitor more than one hybridization at a time, makes this a cumbersome method for most DNA probe applications.

Most other methods require one of the DNA species involved in the hybridization to be labelled (usually radiolabelled) and the sequestering of the labelled DNA into hybrid is monitored. In solution hybridizations, with both reactants in solution, a number of techniques may be used to assess the amount of double-stranded hybrid formed: hydroxyapatite fractionation, S1 nuclease digestion, etc. In solid phase hybridization, however, the target DNA species is bound to a solid support and the labelled DNA, in solution, is allowed to hybridize with the solid-bound DNA. The amount of hybrid formed can simply be determined by washing of the solid support and measurement of the bound radioactivity by scintillation counting, autoradiography, etc. In Southern or Northern hybridization, the solid support is a membrane (usually nitrocellulose or nylon-based) that has been blotted with DNA (or RNA) transferred from a gel in which the target DNA (or RNA) samples have been separated by electrophoresis. Target DNA may also be applied directly to a membrane by filtration, replica plating, or direct application (dot-blotting). In addition other solid supports, such as activated cellulose, polystyrene, or even magnetic beads may also be used.

2 APPLICATION OF DNA HYBRIDIZATION

All DNA probe applications require that DNA be first released from the cells; and for most applications the DNA must then be purified. There are a number of methods available for DNA extraction. We presently use a straightforward enzymic protocol that is outlined in the legend to Figure 7.2. This may not be optimal for all mycobacteria but does give us some DNA from all mycobacteria we have examined. More efficient cell lysis and therefore higher DNA yields, may be achieved if the mycobacteria are grown in the presence of cycloserine and ampicillin (see Crawford and Falkinham, this volume).

2.1 DNA homology determinations

The rate of hybridization between two species of genomic DNA will be proportional to the concentration of homologous DNA and therefore measurement of hybridization kinetics has been widely used to estimate DNA homology between bacterial and mycobacterial species (see Clark-Curtiss, this volume). Hybridizations are normally performed at temperatures 25°C below the T_m, although performing hybridizations at a number of different stringencies will reveal additional information on the fraction of DNA at varying degrees of DNA homology. An additional refinement is to measure the T_m of hybrid formed, since this can be sensitive to low levels of mismatching.

Much of our current understanding of mycobacterial phylogeny has been gained by measurement use of DNA hybridization to measure DNA homology between species (see Clark-Curtiss, this volume), and it remains an essential tool in bacterial taxonomy. The success of the technique is most apparent in studies of those mycobacterial groups, such as the MAI (*Mycobacterium avium-intracellulare*) complex. This group of mycobacteria, comprising the mainly environmental opportunists *M. avium* and *M. intracellulare*, (*Mycobacterium scrofulaceum* is also sometimes included, when the complex is termed the MAIS complex) but also the specific pathogens of ruminants and rodents respectively: *M. paratuberculosis* and *M. lepraemurium*. These mycobacteria are poorly resolved by conventional criteria. The work of Baess (Baess, 1979, 1983, 1984; Baess and Weiss Bentzon, 1978) clearly demonstrated that though *M. avium*, *M. intracellulare* and *M. scrofulaceum*, were very similar phenotypically, they in fact share only 34–56% DNA homology, not much more than they share with dissimilar species such as *M. tuberculosis*. However some strains previously described as *M. intracellulare* were shown to be most homologous to *M. avium*. A number of studies established that *M. paratuberculosis* was indistinguishable from *M. avium* (95–100% DNA homology) by these techniques (McFadden *et al.*, 1987a; Hurley *et al.*, 1988; Saxegaard and Baess, 1988; Yoshimura and Graham, 1988).

Essentially the same techniques may be used to measure the DNA homology between a cloned DNA probe (or even oligonucleotide) and the homologous sequence in target DNA. This may be particularly useful when DNA from one of the species is limited, but cloned DNA is available (e.g. *M. leprae*). It should be remembered however that whereas measurement of hybridization between two samples of total DNA may be used to estimate total DNA homology between two organisms, when a specific DNA probe is used the amount of hybrid formed will be a reflection only of the homology between the probe sequence and the homologous DNA in the target. This will not necessarily be an accurate reflection of the overall level of homology between the target DNA and total DNA from the species from which the probe was derived.

A limitation of the DNA homology determinations is that DNA samples differing by less than approximately 5% in DNA homology cannot be accurately resolved. Therefore, although the technique has very successfully established the coarse taxonomic structure of the genus *Mycobacterium*, it cannot be used to differentiate between very closely related strains (as may be required in epidemiological studies) or species. *M. tuberculosis* and *M. bovis* are 95–100% homologous by this technique, as are *M. avium*, *M. paratuberculosis* and *M. lepraemurium* (Gross and Wayne, 1970; Baess and Weis Bentzon, 1978; Baess, 1979, 1983; Athwal *et al.*, 1984; McFadden *et al.*, 1987a; Saxegaard and Baess, 1988; Yoshimura and Graham, 1988; Hurley, *et al.*, 1988; see Clark-Curtiss, this volume). Current guidelines in bacterial taxonomy are that bacteria with greater than 60–70% DNA homology should be regarded as strains of the same species (Wayne, 1978), leading many workers to suggest rationalization of some of these mycobacterial species names. However most clinical and veterinary microbiologists would not welcome any name changes for these important pathogens and it is still necessary to distinguish these mycobacteria for epidemiological reasons. In any case, it must be remembered that indistinguishable does not mean identical — use of more discriminating DNA probe techniques, as described below, clearly differentiates these pathogens indicating that they are genetically distinct and biologically isolated. Whether this is enough to warrant species status is a question that has yet to be adequately resolved in bacterial taxonomy.

An additional limitation of DNA homology studies is that very low levels of DNA homology are similarly uninformative and therefore the method is not useful for inferring 'long range' phylogenetic relationships. The relationship of *M. leprae* to the cultivatable mycobacteria is confounded by this problem since it shows a similar low level of homology with all mycobacteria so far tested (see Clark-Curtiss, this volume). Sequencing highly conserved genes, such as those encoding ribosomal RNA is the method of choice for inferring long-range phylogenies (see Clark-Curtiss, this volume). In order to detect small differences in DNA sequence between closely related strains and

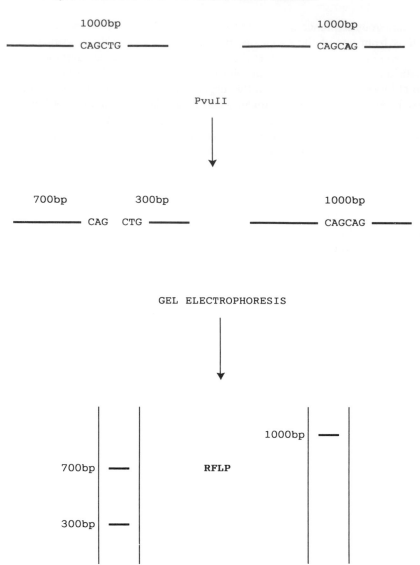

Figure 7.1 Restriction fragment length polymorphism (RFLP). Illustration of how a single base substitution in a 1000bp DNA fragment may be detected as an RFLP.

species, more sensitive methods are required. Clearly the most sensitive is direct DNA sequencing. However, despite enormous advances in the speed and accuracy of DNA sequencing it is not yet a technique that can rapidly be applied to a large number of samples. A more readily applied technique requiring little expertise is the detection of restriction fragment-length polymorphisms (RFLPs).

2.2 RFLP detection

If two dsDNA species of length, say 1000bp are identical at all but one base and if that base difference lies within the recognition site sequence of a restriction endonuclease (RE), then digestion of both samples with that restriction endonuclease will produce different products (Figure 7.1). If the products are resolved by electrophoresis then banding differences will be obtained. These banding differences are usually referred to as restriction fragment-length polymorphisms (RFLPs).

Most restriction endonucleases recognize either a 4bp or 6bp sequence; therefore in any random sequence the frequency of sites will be either approximately every 256bp (4^4) or 4096bp (4^6) respectively, the exact value depending on the relative base composition of restriction target site and DNA sample. For small genomes (e.g. viruses) electrophoresis and staining of restriction fragments produced by digestion allows easy detection of RFLPs. However as the genome size increases, the number of fragments produced increase proportionally, so that for a typical mycobacterial genome of approximately 5 million basepairs somewhere in the region of 1000 fragments would be produced by a typical 6bp restriction endonuclease; and the highly complex banding patterns obtained are then only poorly resolved by electrophoresis. Nevertheless with good-quality DNA and careful electrophoresis on long agarose gels, complex banding patterns may be seen, particularly in the high molecular weight region. These banding patterns, sometimes referred to as genomic fingerprints, may be used to differentiate between closely related species and even within species. Using this technique, *M. tuberculosis* and *M. bovis* were found to give similar but distinguishable genomic fingerprints (Collins and de Lisle, 1984) and clear differences were observed between strains of either *M. bovis* or *M. tuberculosis*. Members of the MAI complex were similarly distinguishable (Patel *et al.*, 1986; Collins and de Lisle, 1986; Whipple *et al.*, 1987), with results in broad agreement with the DNA homology studies described above; however, in addition, *M. paratuberculosis* strains were shown to produce very similar but distinguishable genomic fingerprints to those of *M. avium*.

The direct examination of RE digests of genomic DNA although useful and relatively simple to perform is limited by the resolving power of the gels. The use of pulsed field gel electrophoresis to resolve the limited number of very large DNA fragments generated by REs recognizing rare (e.g. AT-rich) sites in mycobacterial DNA has recently been used to produce simpler and more easily interpreted genomic fingerprints that can be used to differentiate mycobacteria (Levy-Frebault *et al.*, 1989). This technique will be highly useful in large-scale mapping of mycobacterial genomes; however, the requirement for very high molecular weight DNA, costly apparatus and some expertise in performing the electrophoresis, presently inhibits its widescale use.

An alternative method of simplifying complex genomic fingerprints is to use a specific DNA probe to examine only those DNA fragments containing sequences homologous to the probe. DNA samples are digested with an RE and electrophoresed as above. However after electrophoresis the DNA in the gel is denatured and 'Southern blotted' to a nitrocellulose or nylon membrane. The membrane is then hybridized with a labelled DNA probe that will bind to bands containing homologous DNA, on the filter. Visualization of the bound DNA probe (by autoradiography for radiolabelled probes) reveals simple banding patterns (Figure 7.2 shows a number of serotyped MAI complex strains). This technique has a number of advantages over direct visualization of genomic fingerprints. Firstly, RFLPs can be very clearly visualized. Secondly, once the information using one DNA probe has been recorded, the probe can be removed and the membrane hybridized with additional DNA probes. Essentially each (independent) DNA probe can be used to examine a different portion of the mycobacterial genome for RFLPs. Since an almost limitless supply of DNA probes can be generated by genomic cloning, very low levels of base substitution (inverse of DNA homology) can be detected as rare RFLPs between closely related strains and species. Lastly, the technique allows examination of a large number of samples simultaneously.

An additional advantage of the use of DNA probes to detect RFLPs is that the data generated can be used to estimate levels of base substitution between closely related strains and species (Upholt, 1977). If a number of DNA probes (and REs) are used to examine two DNA samples, the number of conserved (homologous) bands that are common between two DNA samples is determined as a fraction (F) of the total number of hybridizing bands. The value of F is a function of the probability of base substitutions occurring within the RE sites located within the target sequence (to generate an RFLP) and therefore will be inversely proportional to the base substitution between the strains. Essentially, each probe examines the restriction enzymes sites in the two DNA samples for conservation of sequence. If DNA base substitution is randomly distributed throughout the genome and the probes used are also from random genomic loci, then the probability of sequence identity at the restriction sites within the sequence homologous to the probes, should follow binomial distribution laws. Therefore the level of base substitution (P) may be expressed as a function of F (Upholt, 1977).

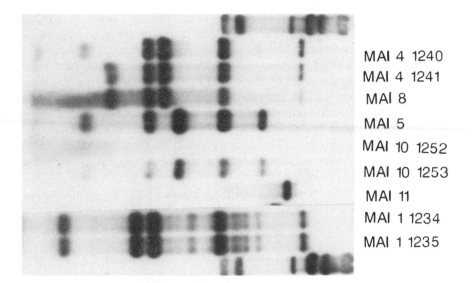

MAI 4 1240
MAI 4 1241
MAI 8
MAI 5
MAI 10 1252
MAI 10 1253
MAI 11
MAI 1 1234
MAI 1 1235

Figure 7.2   RFLP typing of MAI strains. Mycobacteria were grown either on slants (LJ slopes). Cells (approx. 0.1 g) were harvested into 5 ml TEN buffer (50 mM tris-HCl pH8, 100 mM EDTA, 150 mM NaCl), washed, then digested with 10 mg/ml subtilisin (Carlsburg, Sigma type VIII) for three hours at 37°C, followed by 1 mg/ml lysozyme (Sigma) for three hours at 50°C. SDS was added to 1% and pronase (Calbiochem) added to 3 mg/ml and incubated at 37°C for 24 hours, with a further addition of pronase (to a total of 6 mg/ml) after 18 hours. DNA was extracted by multiple phenol/chloroform extraction, ribonuclease A digestion and ethanol precipitation as described (McFadden *et al.*, 1987a). DNA was digested with *PvuII*, electrophoresed through 1% agarose, blotted onto nylon membrane, hybridized to [32P]-radiolabelled, by multiprime labelling (Amersham, UK) pMB22, in hybridization solution containing 0.5 M sodium chloride at 65°C overnight, washed several times in 0.15 M sodium chloride buffer at 65°C and autoradiographed (McFadden *et al.*, 1987b, 1987c). The mycobacterial DNA samples are flanked by Hind III-digested phage lambda DNA.

$$P \approx 1 - [-F + (F^2 + 8F)^{1/2}/2]^{1/n}$$

here n is the number of bases in the restriction enzyme site (usually 4 or 6).

This equation is only valid if there are no gross DNA rearrangements (e.g. transposition) and the probability of multiple substitutions within the restriction endonuclease sites is low. However the equation can be used to estimate the base substitution between DNA samples having greater than approximately 85% DNA homology. The standard error of estimating P may be determined using the equation:

Standard error $\rho = [P(1 - P)/N]^{1/2}$

where N is the total number of bases examined, which is equal to the number of fragments examined, multiplied by the number of bases (n) in the recognition site of the enzyme (McFadden *et al.*, 1987b). Clearly the greater number of bands compared, the lower the standard error. It should be remembered however that the choice of probe may significantly influenne the probability of detecting base changes; i.e. DNA probes from conserved genes (such as the heat-shock genes or ribosomal RNA genes) will show few band differences between strains and conversely for loci containing mutation hotspots. For an accurate estimate of base substitution the best probes to use are randomly-derived clones homologous to single copy DNA showing no evidence of rearrangements. The data obtained may be used to construct phylogenetic trees between closely related strains or species, as shown in Figure 7.3, for MAI complex strains.

Examination of some mycobacterial populations using this technique has revealed that some species or strains appear to be clonal in that very low levels

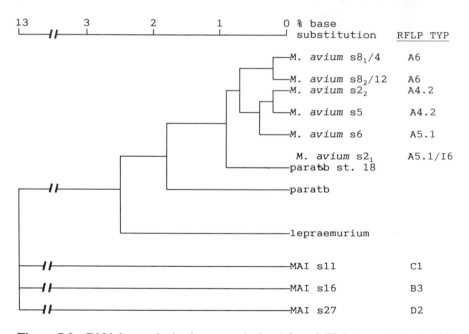

Figure 7.3 DNA base substitutions, as calculated from RFLP data, within the *M. avium* complex depicted in the form of a phylogenetic tree. The horizontal scale corresponds to P, the estimated base substitution. The vertical bar at $P = 13$ corresponds to the approximate point at which the banding patterns are completely different, which is the minimal estimate of the true value of P between organisms connected by this line. The s numbers correspond to MAI serotypes and the subscripts to specific strains, as described by McFadden *et al.* (1987c). RFLP TYP refers to *M. avium* RFLP type.

of base substitution are found between strains (McFadden, 1987b, 1987c). To estimate the confidence limits on the identity of two strains for which no RFLPs were detected, it is necessary to calculate the maximum fraction of base substitution (P_m) compatible with identical banding patterns. The binomial test statistic Z may be used to estimate this value.

$$P_m = Z^2/(N + Z^2)$$

where N is the total number of independent bases examined, which is equal to the number of fragments examined times the number of bases in the recognition sequence of the enzyme producing the fragment pattern (McFadden et al., 1987b). The value of Z is obtained from tables of area under the normal curve: for a level of significance of 95% (95% confidence that base substitutions between two samples is less than P_m), Z is 1.64.

RFLP analysis has since been applied to a number of mycobacterial pathogens and groups, including the TB complex, the MAI complex and M. leprae, as described by Clark-Curtiss (this volume). RFLP typing is useful as either a method of species identification, or as a method for subdividing species into a number of RFLP types. DNA probes have usually been cloned DNA fragments, although one of the earliest uses of DNA probes to detect RFLPs in mycobacterial DNA, by Shoemaker et al. (1986) utilized radio-labelled total genomic DNA from M. tuberculosis to hybridize to genomic digests. One might expect that in this situation the autoradiograph obtained would be essentially identical to a photograph of the gel. However, because the sensitivity of detection is greater, small quantities of relatively large DNA fragments separated from the majority of smaller DNA fragments may be detected. Clear differences in banding patterns were obtained for most strains of M. tuberculosis tested. Indeed, the number of distinctly different banding patterns obtained was surprising considering the low level of base substitution expected between strains of M. tuberculosis. However, in hybridization reactions, obeying second order kinetics, the rate of DNA hybridization is proportional to the square of the homologous probe DNA concentration; therefore repetitive DNA will contribute proportionally more to the signal obtained than single copy DNA (since the relative hybridization rate for repetitive DNA will be proportional to the square of copy number). It is likely therefore that Shoemaker et al. were detecting bands containing polymorphic repetitive elements in the M. tuberculosis genome. Indeed the banding patterns appear similar to those later obtained by other workers using M. tuberculosis-specific repetitive cloned DNA probes, as described below.

The value of RFLP analysis in species identification is most clearly illustrated in examining those mycobacterial species not clearly differentiated by other methods. McFadden et al. (1987c) and Picken et al. (1988) examined the MAI complex and demonstrated that M. avium and M. intracellulare produced completely dissimilar banding patterns that could be easily differentiated, confirming their genetic divergence suggested by DNA homology studies. Many serotypes previously associated with *M. intracellulare* were found to be *M. avium*. A number of RFLP types of each species could be recognized, and strains of *M. intracellulare* were found to be much more heterogeneous than *M. avium*. RFLP typing could also be applied to specific epidemiological questions, e.g. McFadden *et al.* (1987b, 1988) was able to show that *M. avium* and *M. paratuberculosis* could be differentiated by RFLP analysis and that Crohn's disease-isolated mycobacteria were identical to *M. paratuberculosis*. Similarly, Hampson *et al.* (1989) were able to show that MAI strains isolated from AIDS patients were almost all *M. avium*, rather than *M. intracellulare*.

RFLP analysis for strain identification is presently most useful in epidemiological studies, but may also have clinical relevance in identifying previously unrecognized pathogen/disease associations. Hampson *et al.* (1989) were able to show that greater than 70% of MAI strains isolated from AIDS patients belonged to a specific RFLP type of *M. avium*, RFLP type A. As in earlier work (McFadden *et al.*, 1987c), it was found that serotyping did not always correspond to genetic identity, as determined by RFLPs (see e.g. serotype 4 and 8 strains in Figure 7.2). We are presently investigating whether RFLP typing may be useful in predicting mycobacterial virulence and/or drug resistance.

RFLP studies have also been used to estimate base-substitution between closely related strains and species: less than 2% base substitution were found to separate *M. paratuberculosis* and *M. avium* (McFadden *et al.*, 1987b, 1987c); and a similar level of base substitution was found to separate *M. tuberculosis* and *M. bovis* (Cooper *et al.*, 1989). Base substitutions between pairs of strains may be used to construct a phylogenetic tree. Figure 7.3 shows such a tree constructed for members of the MAI complex. Such trees may be useful in order to reconstruct the evolutionary history of mycobacteria and attempt to identify genetic changes that may be responsible for shifts in biological properties. It is for instance interesting that at least two independent sublines that have evolved from *M. avium* (*M. paratuberculosis* and *M. lepraemurium*) are characterized by having complex growth factor requirements and altered pathogenicity compared to *M. avium*. Evolution of one of these (*M. paratuberculosis*) was associated with acquisition of multiple copies of an insertion sequence IS*900* (see below and Martin *et al.*, this volume). It would certainly be of interest to determine the evolutionary history of *M. leprae* in this way; however, so far no close relatives of *M. leprae* have been found (see Clark-Curtiss, this volume) and this remains a challenge for DNA probes.

For very closely related strains within a species, the level of base substitution may be very low and therefore informative RFLPs may be difficult to detect. Cooper *et al.* (1989) were unable to distinguish *M. bovis* strains using random genomic clones as probes. Similarly, most strains of *M. leprae* were indistinguishable by this technique (see Clark-Curtiss, this volume). A strategy to obtain more informative probes would be to identify polymorphic regions of the genome. These may be plasmids, integrated phages, hot-spots for mutation or recombination, transposons or elements of unknown function. A common feature of many of these elements is that they are present in multiple copy in the genome. The use of plasmids as DNA probes to examine the MAIS complex is discussed by Crawford and Falkinham in this volume. Clones containing repetitive DNA may be detected by screening randomly isolated clones for any that hybridize to an anomalous large number of bands on Southern blots. Alternatively, screening a genomic DNA library with labelled total parent DNA (as described for the Shoemaker *et al.* experiments above) should identify clones containing

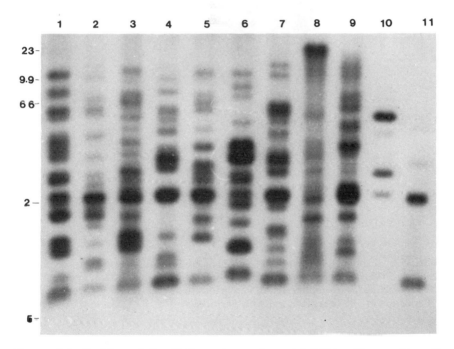

Figure 7.4 Probing of *PvuII*-digested mycobacterial DNA with repetitive *M. tuberculosis* probe A3/2. Tracks 1–9 show various independently derived *M. tuberculosis* strains, lane 10 *M. bovis*, lane 11 *M. bovis* BCG. Molecular size markers (kilobases) are shown on the left. DNA extractions, electrophoresis and hybridizations are as described (Zainuddin and Dale, 1989). From Zainuddin and Dale, (1989) with permission.

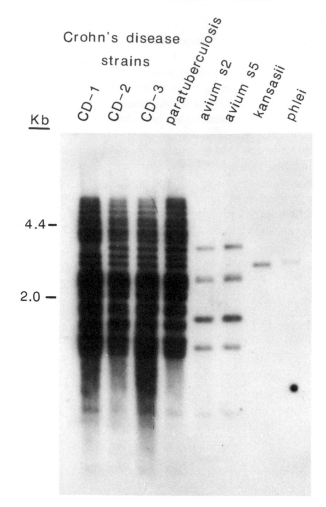

Figure 7.5 Probing of *PvuII*-digested mycobacterial species DNA, as shown, including Crohn's disease-derived strains, CD-1, CD-2 and CD-3, with pMB22. DNA extractions, electrophoresis and hybridizations were as in Figure 7.2. From McFadden *et al.*, 1988 reproduced with permission.

repetitive elements. A number of groups have isolated DNA probes containing repetitive elements from *M. tuberculosis* (Eisenach *et al.*, 1988; Reddi *et al.*, 1988; Zainuddin and Dale, 1989; see Clark-Curtiss, this volume). At least some of these appear to be mycobacterial insertion sequences (Thierry *et al.*, 1990; see Martin *et al.*, this volume) and give polymorphic banding patterns. Figure 7.4 shows a number of *M. tuberculosis* strains, *M. bovis* and *M. bovis*

BCG, probed with the repetitive element clone A3/2 (Zainuddin and Dale, 1989). Highly polymorphic bands are obtained, particularly for *M. tuberculosis* strains; and *M. bovis* can be clearly distinguished from both BCG and *M. tuberculosis*. Probes containing these elements should therefore be of considerable value in epidemiological studies. In contrast, repetitive DNA elements from *M. paratuberculosis* and *M. leprae* (Clark-Curtiss and Docherty, 1988; Grosskinsky *et al.*, 1989; see Clark-Curtiss, this volume) were found to produce identical or very similar banding patterns for almost all strains of each species. The value of these elements is however their species specificity. Figure 7.5 shows a group of Crohn's-isolated mycobacteria and classified mycobacteria probed with the clone pMB22, which contains a single copy of the element IS900, which is specific for *M. paratuberculosis* (Green *et al.*, 1989; McFadden *et al.*, 1990; see Martin *et al.*, this volume). Using this probe, multiple near-identical banding patterns are obtained for all *M. paratuberculosis* strains, including those derived from Crohn's disease tissues (McFadden *et al.*, 1987b, 1987c, 1988, 1990). Repetitive elements that are species-specific, will of course be also very useful as probes for specific detection (discussed below).

Most RFLP typing on mycobacteria has been performed using radio-labelled DNA probes. Handling and disposal problems coupled with short shelf life has inhibited the adoption of this technique beyond research laboratories. However, techniques presently available for non-isotopic labelling of DNA are simple to perform, similar in sensitivity to radiolabelling with e.g. ^{32}P, and the probes are stable. Figure 7.6B shows the result obtained, hybridizing mycobacterial DNA, mostly MAI strains, with a probe labelled with digoxygenin modified nucleotides. The bound probe is detected with enzyme–antibody conjugate and chromogenic substrate. The result is comparable with that obtained when the same filter was hybridized with conventionally radiolabelled probe (Figure 7.6A).

Examination of Figures 7.2–7.6 illustrates how we presently RFLP type *Pvu*II-digested mycobacterial DNA using the clone pMB22. *M. paratuberculosis* (as in Figure 7.5) gives a highly complex but almost invariant banding pattern with pMB22, due to hybridization of the copy of the insertion sequence IS900 in pMB22, to multiple genomic copies of IS900 in its genome. The simpler and more variable *M. avium* banding pattern seen in the same figure reflects hybridization with genomic DNA flanking the copy of IS900 present in pMB22. The banding pattern given by the MAI serotype 4, strain 1241 (Figure 7.2), we designate as *M. avium* RFLP type A6 (type A, 6 bands — RFLP type A of Hampson *et al.*, 1989) and related banding patterns are designated e.g. A4.2 (4 bands in common with A6, two unique), etc. More distantly related mycobacteria (greater than 13% base substitution) including those, such as *M. intracellulare* strains that are part of the MAI

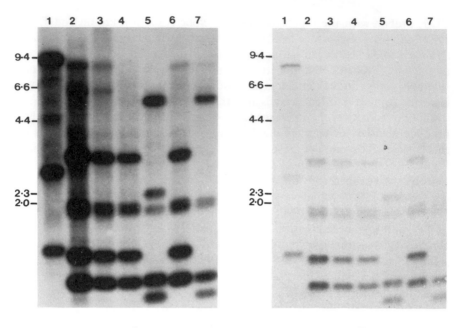

Figure 7.6   Probing of MAI strains with radiolabelled and non-isotopic probes. (A) DNA was extracted from seven MAI strains, as in Figure 7.2, probed with ^{32}P-radiolabelled pMB22 and autoradiographed for six hours. (B) Probe was removed from the membrane by washing in alkali and the filter then hybridized to pMB22 that was labelled by incorporation of the nucleotide analogue, digoxigenin-11-dUTP, by random primed labelling (Boehringer Mannheim, Mannheim, W. Germany). The bound probe was detected with anti-digoxigenin-alkaline-phosphatase enzyme conjugated antibody. The location of the antibody–antigen conjugate on the filter was detected by using a chromogenic substrate and photographed (B). Molecular size markers (Kb) are shown on the left.

complex give completely dissimilar patterns (e.g. MAI 11 in Figure 7.2, lane 1 of Figure 7.6 and MAI serotypes 11, 16 and 27 in Figure 7.3), and are designated e.g. *M. avium* complex RFLP type C*1*. Note from Figure 7.3, that strains of the same (pMB22) RFLP type are closely related but not necessarily genetically identical — additional probes may differentiate them. Some *M. avium* strains, such as the two MAI serotype 1 strains in Figure 7.2, *M. avium* 2_1 and *M. paratuberculosis* strain 18 of Figure 7.3, give a group of additional bands when probed with pMB22, that are not present in closely related *M. avium* strains (e.g. MAI 4 strain 1240, Figure 7.2). We have demonstrated that these bands are due to the presence of repetitive elements related to IS*900*, present in these strains (McFadden *et al.*,

1990); and we designate these strains as e.g. *M. avium* RFLP type *A5.1/I6.*

It is interesting to examine the population structure of mycobacterial pathogens using RFLPs. The very high degree of sequence conservation of the strict pathogens *M. paratuberculosis* and *M. leprae* has already been noted (see Clark-Curtiss, this volume). However for other mycobacterial species (*M. avium*, *M. tuberculosis*) the variation in RFLP pattern appears also to be limited. We and others have now examined hundreds of *M. avium* complex strains from a variety of sources and most fall into a very limited number of highly conserved RFLP types (McFadden *et al.* 1987c, 1988; Picken *et al.*, 1988; Hampson *et al.*, 1989 and our unpublished results). *M. bovis* seems similarly invariant (Cooper *et al.*, 1989); and although repetitive probes from *M. tuberculosis* detect polymorphic repetitive elements, random genomic probes detect a much lower degree of heterogeneity (Eisenach *et al.*, 1986). These results may indicate that mycobacterial populations are clonal in structure and that the limited amount of genetic variation found may be largely associated with DNA transposition. New variants that arise by mutations within a population may replace, partly or wholly, the present incumbants of a particular ecological niche (e.g. ruminants susceptible to Johne's disease or humans susceptible to leprosy), leading to only a limited amount of genetic variation present in that population at any one time.

2.3 DNA probes for detection

An important reason why DNA probe technology has been slow to be applied in diagnostic microbiological laboratories is the success of the conventional technology. More than 100 years of refinement has contributed to the simplicity, speed, sensitivity and economy of conventional diagnostic microbiology. With little more than petri dishes, media, a bunsen burner and a few simple biochemical tests, a microbiologist may detect and identify most clinically significant bacteria in less than 24 hours. However, this is manifestly not the case in the mycobacterial diagnostic laboratory. Growth of *M. tuberculosis* to single colonies takes at least two weeks. Other mycobacteria, such as *M. paratuberculosis* or *Mycobacterium ulcerans* take much longer, 6–12 weeks. And then of course there is the leprosy bacillus; which despite over a century of ingenious efforts and numerous false dawns, has resisted all attempts to persuade it to replicate *in vitro*. It is therefore not surprising then that mycobacterial laboratories have been amongst the first to utilize DNA probes for detection.

DNA probes have a number of advantages over protein-based (e.g. immunological) diagnostic methods. The stability of DNA probes and their targets (as evidenced recently by the amplification of DNA from extinct animals and a 7000 year-old human), as compared to protein-based systems has advantages in transport and storage of reagents and samples, particularly in the developing world where the maintenance of cold chains remains a considerable logistic problem (Daniel, 1989). Manufacture of DNA probes on a commercial scale is much more easily attained than that of monoclonal antibodies or other protein reagents. The specificity of DNA probes, with their theoretical ability to differentiate two organisms differing in only a single base substitution, may be useful in differentiating closely related mycobacteria. Lastly, with the introduction of the polymerase chain reaction (PCR to be described below), DNA probes have the capability of exquisite sensitivity that is unlikely to be matched by immunological assays, allowing direct application to clinical specimens infected with cultivatable or non-cultivatable mycobacteria.

To evaluate the possible advantages of DNA probe technology, it is useful to consider first the demands of the diagnostic mycobacterial laboratory. Clinical or veterinary specimens are presented to the laboratory as a suspension of bacteria within a body fluid, such as sputum, gastric aspirate, urine, CSF or faeces; or the specimen may be a sample of tissue, such as a lymph node. In Figure 7.7, the typical microbiological investigations used to examine respiratory specimens, such as sputum, from patients with suspected tuberculosis, are plotted against time. The precise procedures followed and times taken vary greatly from laboratory to laboratory and will also depend on the number of organisms present in the original inoculum; however the general scheme, as presented in Figure 7.7, is a reasonable approximation of what happens in most laboratories. The first investigation is invariable microscopic detection of acid-fast bacilli. Although simple, rapid and economical, the estimated detection limit of direct microscopy is approximately 10^4 bacilli per millilitre of sputum (Yaeger et $al.$, 1966), considerably less than culture and therefore most culture positive specimens are sputum negative. In addition, the presence of saprophytic mycobacteria in some specimens (e.g. gastric aspirates, faeces) may give rise to false positives. For definitive diagnosis of TB, culture is presently always required. There are two main cultural alternatives presently available to diagnostic laboratories: conventional culture on media, such as Loewenstein-Jensen (LJ) slopes (unbroken lines in Figure 7.7), and the more rapid radiometric culture in liquid media using the 'BACTEC' system (dotted lines in Figure 7.7 — Roberts et $al.$, 1983). The time for culture positivity (CLTR POS on Figure 7.7) differs for specimens that are positive and negative by microscopy as indicated for both procedures. The maximum sensitivity of the two systems is thought to be similar at about 10^2 bacilli per millilitre. Positive cultures are further investigated by a specialist laboratory: usually species identification (IDENT on figure) by biochemical tests, and drug sensitivity

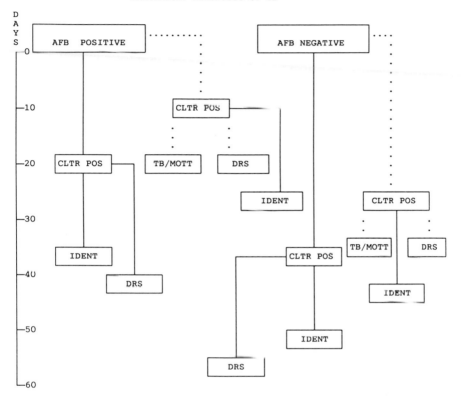

Figure 7.7   Schematic representation of the timetable of microbiological procedures involved in examining specimens, such as sputa. Specimens are first examined for the presence of acid-fast bacilli (AFB) and may be positive or negative. The sample is then inoculated, either into conventional solid media (solid lines) or radiometric (BACTEC – dotted lines) culture. Presence of a visible colony of acid-fast bacteria on solid media, or a positive radiometric reading on the BACTEC system is indicated as CLTR POS. Samples that are culture positive on solid media are then further investigated, usually by a specialist laboratory for species identification (IDENT) by biochemical tests, and drug sensitivity determinations (DRS). Positive radiometric cultures from the BACTEC system may be processed further by subculture (onto solid media) and standard biochemical species identification. However, the BACTEC system allows a more rapid differentiation of positive cultures into either TB complex or mycobacteria other than TB complex (TB/MOTT), on the basis of growth in the presence of NAP (*p*-nitro-acetylamino-hydroxypropiophenone); and also allows more rapid determination of drug sensitivities.

determinations — (DRS on figure), are performed. Positive radiometric cultures from the BACTEC system may be processed further by subculture (onto e.g. LJ slopes) and standard biochemical species identification. However, the BACTEC system allows a more rapid differentiation of positive cultures into either TB complex or mycobacteria other than TB complex (MOTT), on the basis of growth in the presence of NAP (p-nitro-acetylamino-hydroxypropiophenone), and also allows rapid determination of drug sensitivities. The microbiological investigations provide information at a number of levels. Firstly, detection of any *Mycobacterium*: this may be from finding of acid-fast bacilli in a smear, a mycobacterial colony on LJ slope or a positive radiometric culture reading. Any of these may be considered enough to confirm a diagnosis of tuberculosis. Indeed, clinical diagnosis of tuberculosis and initiation of drug therapy may often be made before any microbiological results, solely on the basis of clinical and radiological findings. However, diagnosis must always be confirmed by culture and species identification must still be obtained for a number of reasons. Firstly, mycobacteria other than *M. tuberculosis* complex (often termed simply MOTT), particularly MAI complex strains, that are commonly isolated from AIDS patients, are often resistant to the drugs normally prescribed for *M. tuberculosis*, and require alternative treatments (e.g. surgical removal of affected tissue and/or additional drugs). Secondly, isolation of *M. tuberculosis* will usually initiate laborious and expensive contact tracing and investigations, whereas isolation of the 'non-transmissible' mycobacteria (*M. bovis*, MOTT) does not necessitate such measures. Contact tracing is often initiated on clinical diagnosis of TB and therefore the more rapidly species identification is made, the sooner unnecessary contact tracing may be abandoned. Drug sensitivities also require culture; however, the number of drug-resistant *M. tuberculosis* is still very low, at least in the west, and species identification (for MOTT, particularly MAI) may be sufficient to predict optimal drug therapy.

The sensitivity of culture is very high for most respiratory specimens such as sputum. However, diagnosis of mycobacterial disease, particularly extra-pulmonary disease, may still be made on clinical findings, despite negative culture results. This is especially true with some specimens such as CSF from patients with suspected tuberculous meningitis or tissue biopsies from patients with suspected tuberculous lymphadenitis, where the sensitivity of culture is thought to be low. It will be a challenge for DNA probes to detect and identify mycobacteria in these specimens.

DNA probes should provide mycobacterial detection and identification, considerably more rapidly than available by conventional culture. DNA probes may also provide greater sensitivity than conventional culture. For leprosy, of course, there is no possibility of culture and DNA probes would clearly be of value. The point at which DNA probes could be applied will depend on the sensitivity of the probing technique. However, the information content of DNA probes may be restricted in that a single probe or combination of probes may identify mycobacteria at any of a number of levels: firstly as a member of the genus *Mycobacterium*; secondly, as a member of a complex such as the TB or MAI complex; thirdly as a specific mycobacterial species, such as *M. avium* or *M. bovis*, or lastly, as a subspecies or strain, such as *M. avium* RFLP type A or a strain carrying a drug resistance gene. If information additional to that obtained from use of a specific DNA probe, is required (e.g. drug sensitivities), then conventional culture may still be necessary.

Diagnostic tests are usually evaluated in terms of their sensitivity (percentage of positive specimens tested positive) and specificity (percentage of negative specimens tested negative). From these values the predictive accuracy of positive or negative test results may be calculated for given prevalence rates of disease in the population (of clinical samples) to be studied. For the clinician, the predictive value of positive tests is most important, since false-positives may lead to inappropriate therapy; whereas false-negatives will usually lead to the further investigations, including usually, repeat testing. As discussed by Daniel (1989), specificity is critical to the predictive accuracy of positive tests, particularly in sample populations where tuberculosis prevalence ranges from 0.15–0.4 (as in TB clinics in most developing countries).

The choice of target DNA sequence for a DNA probe will determine both the sensitivity and specificity. A probe may be the whole mycobacterial genome or some fraction of it (e.g. plasmid), a cloned DNA fragment or a synthetic oligonucleotide. Probes may be derived from DNA of a single mycobacterial species, or be a mixture of DNA sequences from several mycobacteria. The degree of sequence conservation of the target DNA locus will of course have considerable influence on the probe specificity. The nature of the probe will also influence specificity. If a small oligonucleotide is used (e.g. 15–30bp), then small differences in DNA sequence between the probe and the target sequence, in different strains or species, may dramatically affect hybrid stability, and therefore the specificity of oligonucleotide probes is usually high. Hybrid stability with larger DNA probes will usually be less sensitive to small differences in DNA sequence. The choice of target DNA sequence for probe is therefore the most crucial factor in considering probe specificity for cloned probes. The sensitivity of the probe is determined by the detection system and the copy number of the target sequence in the genome. For greatest sensitivity, probe homologous to targets present in multiple copy per cell, such as repetitive DNA or ribosomal RNA, may be used.

Some of the earliest uses of DNA probes for detection of mycobacteria, used radiolabelled whole chromosomal DNA probe from *M. tuberculosis* to demonstrate specific detection of *M. tuberculosis* cell suspensions (Shoemaker *et al.*, 1985). DNA equivalent to 2×10^4 genomes, was detected by a dot-blot assay. The probe was shown to be mycobacterial specific (did not detect other bacteria) but not species specific (crossreacted with MAI DNA). Roberts *et al.* (1987) used whole chromosomal DNA probes and a dot-blot assay specifically to detect and identify both *M. tuberculosis* complex (probe was a whole chromosomal DNA from *M. bovis*) and MAIS complex DNA (probe was a mixture of chromosomal DNA from *M. avium*, *M. intracellulare* and *M. scrofulaceum*) in conventional culture and positive radiometric cultural isolates from the BACTEC system (TB/MOTT point of Figure 7.6). The dot-blot test took up to 48 hours to perform and correctly identified 93% and 59% of cultures from the conventional and BACTEC systems respectively. The percentage correctly identified increased to 100% and 93% respectively when the cultures were treated with antibiotic before probe application in order to obtain more efficient cell lysis. This illustrates an important point in DNA-probe applications; that it is essential to ensure efficient lysis of the cells in order that the target nucleic acid is available for hybridization. A similar slot-blot system, but using cloned DNA probes from *M. tuberculosis* was used to demonstrate detection of *M. tuberculosis* DNA equivalent to 10^4 genomes (Pao *et al.*, 1988). Applying the technique to detect *M. tuberculosis* DNA in sputum samples, a sensitivity of 90% was obtained. This high sensitivity was perhaps surprising since their reported detection limit of 10^4 bacilli is considerably less than that generally estimated for culture. However, a number of false positives were also obtained, giving a specificity of 83% and a predictive accuracy of a positive test of only 56%. Butcher *et al.* (1988) used cloned mycobacterial DNA probes and Southern blotting to attempt to detect mycobacteria DNA in Crohn's disease tissues. No mycobacterial sequences were detected at a sensitivity of less than one mycobacterial genome per 10 human cells. However, (unpublished) attempts to detect *M. tuberculosis* DNA in human tuberculous cervical lymph nodes by this technique were similarly unsuccessful, suggesting that very low numbers of mycobacteria may be present in diseased tissues. A dot-blot assay has also been evaluated for detection of *M. paratuberculosis* in faeces of animals with Johne's disease. Using a cloned DNA probe, Hurley *et al.* (1989) could detect 34% more mycobacterium-containing samples than culture. However, their probe also hybridized to other MAIS complex strains such as *M. avium*; so it was not clear which species was responsible for the greater positivity.

Mycobacterial DNA probes have been developed by both Syngene Co. (Molecular Biosystems Inc., San Diego, CA) and Gen-Probe. However, use of only the latter probes has yet been extensively documented and will be described.

2.3.1 The Gen-Probe rapid diagnostic system

The first DNA probes to be commercially available for mycobacterial detection were those marketed by the company Gen-Probe (Gen-Probe Inc., San Diego, California, USA). Three mycobacterial diagnostic systems are presently available: one for the TB complex and another for the *M. avium* complex (comprising *M. avium* and *M. intracellulare*, but not including *M. scrofulaceum*) and a *Mycobacterium gordonae* test. All systems presently use ^{125}I-labelled single-stranded synthetic oligonucleotides complementary to specific ribosomal RNA sequences (rRNA), as probes. Since rRNA is present in approximately 10 000 copies per bacterial cell, the use of rRNA as target should increase sensitivity of detection. The sequence of rRNA is highly conserved, relative to most other genes, and therefore may be considered an unlikely choice for developing species-specific probes. However, species-specific differences in sequence are found amongst bacteria, particularly in some 'variable' regions of rRNA structure (Woose, 1987). As described above, small differences in sequence are likely to affect hybrid stability of short oligonucleotide probes; therefore it seems likely that the Gen-Probe probes are complementary to unique species-specific sequences of rRNA. The TB complex and the *M. gordonae* probe each consist of a single DNA probe; whereas the *M. avium* complex probe comprises two probes: one specific for *M. avium* and the other specific for *M. intracellulare*. Each system is used to rapidly identify cultures and therefore would be used at any of the points labelled CTLR POS in Figure 7.7.

The procedure for each probe is similar. The starting point is a visible mycobacterial colony, or positive radiometric culture (CLTR POS in Figure 7.7), that must be resuspended to give a turbidity equivalent to a No. 1 McFarland standard, corresponding to approximately 60 million bacteria per millilitre. If too few bacteria are used, insufficient nucleic acid will be available to bind the probe and a false negative result will be obtained. Sherman *et al.* (1989) suggest that cultures are ready for identification when colonies are large enough to be morphologically characterized macroscopically, corresponding to approximately 2–3 weeks from inoculation. The bacterial suspension is then lysed by sonication in a lysing solution. The ^{125}I-labelled probe is then added and allowed to hybridize with any mycobacterial RNA released from the cells. A separation suspension is then used, containing the absorption agent hydroxyapatite which binds DNA:RNA hybrid. The hydroxyapatite-bound probe is collected by centrifugation and uncomplexed probe is removed by washing. Hybridized probe is quantitated using a gamma counter and the result is expressed as a per cent of the input probe hybridized. A per cent hybridization greater or equal to 10% is recommended by the manufacturers as indicating a positive result (although see below). The manufacturers recommend that a positive and negative control be used in each test series. The whole procedure takes approximately two hours. The Gen-Probe system has also been successfully used to identify mycobacteria from radiometric BACTEC cultures. The combination of radiometric culture and the DNA probe cuts time required for identification from 9 to 11 weeks for conventional culture and biochemical identification, to 2–4 weeks.

The ability of the Gen-Probe rapid diagnostic systems to identify correctly mycobacteria, both from conventional or radiometric culture, of the *M. tuberculosis* or *M. avium* complex, has been evaluated by a number of laboratories and shown to have both very high specificity and sensitivity ranging from 95 to 100% (Drake *et al.*, 1987; Kiehn and Edwards, 1987; Gonzalez and Hanna, 1987; Ellner *et al.*, 1988; Musial *et al.*, 1988; Sherman *et al.*, 1989). Where discrepancies have arisen they were usually with samples that gave hybridization values in the range of 5–15%, or, in the case of the *M. avium* complex probes, were sometimes strains that hybridized to both the *M. avium* and the *M. intracellulare* probes. In some cases the discrepancies were due to insufficient sample being used for analysis. Many of these discrepancies were resolved on repeat testing and therefore Drake *et al.* (1987) and Sherman *et al.* (1989) recommend that all samples that give hybridization values in the range of 5–10%, or 5–15% respectively be repeated. In the series of experiments reported by Sherman *et al.* (1989), 73 of 589 clinical specimen cultures that were examined with the DNA probe procedure fell within the 5–15% range; and the authors recommend that only hybridization values greater than 15% (rather than the manufacturer's recommended 10%) be considered positive. A few strains persisted in giving anomalous results and conventional biochemical testing is recommended for cultures that repeatedly give hybridization values in the intermediate range. Mixed cultures were sometimes the cause of strains reacting with both probes; though a few single *M. avium* complex strains consistently hybridized to both the *M. avium* and *M. intracellulare* probe. It is however often very difficult to separate mixed mycobacterial cultures and differentiating these species is not important for clinical evaluation. In the study of Musial *et al.* (1988) two (of 102 tested) *M. tuberculosis* strains consistently hybridized to the *M. intracellulare* probe, in addition to the *M. tuberculosis* probe. Also, occasionally rare organisms are found that do not react with the probes. Sherman *et al.* (1989) report that two isolates, identified as *M. avium* by high-performance liquid chromatography, did not react with either of the MAI probes. Nevertheless, despite these occasional discrepant results, the Gen-Probe tests give results of very high predictive value. In the series of experiments reported by Musial *et al.* (1989), the predictive values for positive and negative tests were 98.0 and 99.6% respectively for the *M. tuberculosis* test and 92.5 and 97.0% respectively for the MAI test.

The Gen-Probe system does not differentiate between *M. tuberculosis* and *M. bovis* (or BCG). This last point has recently been emphasized by Heifets (1989), who considers that isolates identified as *M. tuberculosis* complex by the Gen-Probe system should also be further identified by conventional biochemical testing. However, Ellner *et al.* argue, in a reply to Heifets letter, that in US laboratories the probability of an isolate being *M. bovis* (or BCG) is very low (less than 1%) and identification of an isolate as *M. tuberculosis* complex is sufficient.

The Gen-Probe system represents a considerable advance in rapidly identifying mycobacterial cultures, representing a saving of 3–8 weeks for definitive identification. However, the use of radionucleotides and the short shelf life (one month) may limit its application to busy laboratories licensed for radioactive work. Gen-Probe inform us that new non-isotopic chemiluminescent systems will be available in early 1990 at lower cost than the isotopic system. The present cost of the test system is $200 for a test kit able to test a maximum of 18 samples plus two controls. If 18 samples are tested simultaneously with both the *M. tuberculosis* (one kit) and the MAI test kits (requiring two kits), then the cost of the tests is $33.33 per isolate, not including labour, equipment, etc. The costs could be reduced if the tests were run consecutively, e.g. *M. tuberculosis* test first followed by MAI test on those cultures negative by the *M. tuberculosis* test. Musial *et al.* also note that differentiation of the MAI complex is not important for determining optimal therapy and suggest that it would be more cost effective to combine the *M. avium* and *M. intracellulare* probes into one pool.

The probes are also excellent epidemiological tools. Saito *et al.* (1989) utilized the ability of the Gen-Probe system to differentiate *M. avium* and *M. intracellulare*, to examine the distribution of these species in human and environmental samples from Japan. Considerable variation was found in the ratio of *M. avium*/*M. intracellulare* from different areas and there was some correlation between their prevalence in environmental and clinical samples.

The use of the probes as taxonomic tools is more problematic. The finding of a small number of mycobacterial isolates that give anomalous test results has already been noted. It would be expected that within a population occasional strains would be encountered that have single base mutations at the rRNA target locus. The rate of sequence change in ribosomal RNA is very low, relative to other genes in the genome; however, within any population, strains with rare base substitutions may be present. When examined with the Gen-Probe system, such strains may give either an anomalous positive result with the wrong probe or an anomalous negative result with the correct probe. The success of the Gen-Probe systems suggests that with the mycobacterial pathogens examined, this must occur very infrequently. However, the small number of strains that gave consistently anomalous result may be due to this expected low rate of variation in rRNA sequence within a species. Also, as described above, results of RFLP typing suggest that *M. intracellulare* may be more heterogeneous than previously recognized; possibly accounting for some anomalous results within the MAI complex. It would be useful to examine those mycobacteria giving anomalous results by RFLP typing, in order to evaluate these possibilities. Because of the sensitivity of hybrid stability to sequence changes caused by low levels of intraspecies sequence variation, any taxonomic system based on the hybridization of short oligonucleotides to a single target would occasionally fall foul of rare sequence variants; as indeed would any taxonomic system based on the result of a single test. For this reason, it is our belief that the Gen-Probe system (or indeed any single test result) should not be used as a definitive taxonomic tool; but must be used in combination with other tests. It is however fair to say that Gen-Probe strongly disagree with this view.

2.3.2   The polymerase chain reaction

Ideally, a DNA probe assay would detect and identify a *Mycobacterium* directly in a clinical sample, with the sensitivity of culture, but in a matter of a few hours rather than weeks. The sensitivity of culture is theoretically capable of detecting a single micro-organism in a sample; however in practice the limit is usually not more than 10^2 bacilli/ml. The sensitivity of direct DNA probe detection is limited by the specific activity of the label that can be incorporated in a probe. Present radiolabelling techniques are capable of labelling DNA to specific activity of approximately 10^9 cpm/µg of DNA. For a 1000bp probe 5.4×10^{-4} cpm would be incorporated in each ssDNA molecule. At least 100 cpm would be needed to give a reasonably detectable signal, corresponding to approximately 180 000 probe molecules, or 0.1 pg of DNA. Since an equivalent number of target molecules, and for a single copy gene, a similar number of cells (i.e. 10^4–10^5), must be present to fix this amount of probe, then this sets a lower limit on the sensitivity attainable by direct DNA probing. For targets that are present in higher copy number, these figures will be more favourable, but the requirement in the Gen-Probe system for approximately 3×10^7 bacilli in each test, indicates that the use of rRNA as target does not alone provide the required sensitivity for direct detection. A DNA probe test, based on the Gen-Probe system, but for direct detection, was recently withdrawn due to low sensitivity. Clearly, to achieve higher levels of detection, comparable to or greater than those achievable by culture, either the target must be concentrated (for limited amounts of sample this will not provide much increase in sensitivity) or the target nucleic acid must first be amplified. This latter strategy can now be achieved by enzymic amplification of DNA by the polymerase chain reaction (PCR).

The development of the PCR (Saiki *et al.*, 1985, 1988) at Cetus Corporation, California, USA, looks set to revolutionize DNA probe technology. The technique provides exquisite sensitivity, at least equal and usually greater than that provided by culture, combined with the speed and specificity of DNA probe assays. In addition the PCR is remarkably simple to perform. The PCR provides an analogous amplification to that provided by culture, but instead of the bacterium amplifying itself by replication, a fragment of the bacterial DNA is rapidly amplified by successive rounds of DNA replication utilizing the enzyme DNA polymerase (Figure 7.8). Target DNA is first denatured by heating to 95°C. DNA oligonucleotides, usually 15–25 bases homologous to sequences 50–2000 bases apart on the target DNA, on opposite strands, are added and allowed to anneal to the target DNA. This step — essentially a DNA hybridization, governed by all the factors discussed above — provides the specificity of the reaction. The annealed oligonucleotides are then allowed to prime DNA synthesis by DNA polymerase, synthesizing a copy of each of the strands from the primers. DNA between the primers is effectively duplicated during this step. The DNA is then denatured by heating, the primers allowed to anneal again and the cycle repeated. It can be seen that on each cycle the amount of DNA between the primers is doubled, allowing an exponential increase in the quantity of this DNA. Simple calculations predict that 25 cycles should provide a (2^{25}) 33 million-fold amplification. However, saturation of enzyme, annealing of template and other factors limits amplification. Nevertheless, 1 million-fold amplification of a single copy of a dsDNA molecule of size 300bp will yield approximately 0.3 pg of DNA, a quantity that is detectable by DNA hybridization. With the introduction of thermostable DNA polymerase from *Thermus aquaticus*, the reaction is easily automated, may be performed and analysed in a few hours and therefore provides a rapid and powerful amplification step for DNA detection and analysis.

Like all DNA probe techniques, the PCR requires that the mycobacterial cells are first lysed, in order to release DNA for amplification. Standard enzymic DNA extractions from mycobacterial cells may be used (e.g. Hartskeerl *et al.* 1989). Alternatively, since PCR does not require high molecular weight native DNA, harsher procedures may be used to lyse cells than what would normally be used in DNA extractions. Brissons-Noel *et al.* (1989) used treatment with alkali (0.1 M sodium hydroxide, 2 M sodium chloride, 0.5% sodium dodecylsulphate, 15 minutes, 95°C), followed by phenol/chloroform extraction and ethanol precipitation; whereas Vary *et al.* (1990) use treatment with 0.2 M sodium hydroxide at 120°C. For rapid identification of mycobacterial colonies, we and others have found that a colony may be picked directly into 0.5 ml of water, boiled, centrifuged and 5 µl of the supernatant analysed directly by PCR.

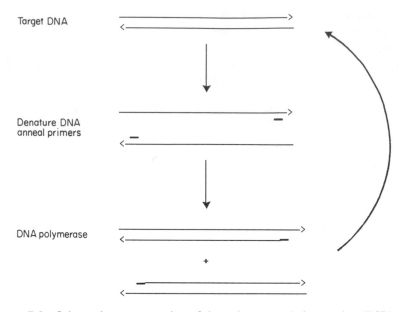

Figure 7.8   Schematic representation of the polymerase chain reaction (PCR).

PCR product may be detected directly by gel electrophoresis and ethidium bromide staining (if greater than approximately 10 ng of product is produced), or by hybridization with labelled probe. Alternatively, a second 'nested' PCR may be used to detect small quantities of PCR product, increasing both sensitivity (about 1000x) and specificity (due to second pair of primers). Figure 7.9 shows a dilution series of *M. paratuberculosis* DNA detected using PCR and primers to the DNA sequence of IS*900* (Green *et al.*, 1989). Panel A shows a standard PCR reaction amplifying a 400bp product — the maximum sensitivity is about 10 pg. In the lower panel 1 µl of the product from the standard reaction has been added to a second, nested PCR reaction, amplifying a 300bp product internal to the 400bp product. Detection of visible product from amplification of 10 fg, approximately two genome equivalents of *M. paratuberculosis* DNA, is easily achieved. We have recently found that the nested reaction may be performed by simply adding the fresh (nested) PCR mixture to frozen primary reaction, minimizing handling and danger of cross-contamination.

The application of the PCR to mycobacterial detection requires DNA sequence information for design and synthesis of primers. Hance *et al.* (1989) used primers homologous to conserved regions of the mycobacterial 65K gene to amplify and sequence a 383bp fragment of the 65K gene in *M. tuberculosis*, *M. bovis*, *M. avium*, *M. paratuberculosis* and *M. fortuitum*.

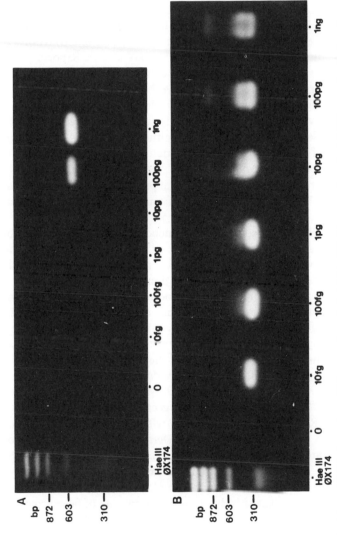

Figure 7.9 Nested PCR detection of *M. paratuberculosis* DNA. (A) *M. paratuberculosis* DNA samples (0–1 ng) were subjected to standard PCR reactions (Saiki *et al.*, 1988) using, IS*900*-specific oligonucleotide primers to amplify a 500bp product, thermostable Taq DNA polymerase in a 50 μl reaction with 30 cycles of: annealing at 68°C for 15 seconds, extension at 72°C for two minutes and denaturing at 94°C for 30 seconds. Some 40 μl of product were then electrophoresed and stained with ethidium bromide. (B) 1 μl of the reaction product from (A) was used in a 30-cycle nested PCR reaction using IS*900*-specific oligonucleotide primers, internal to the primers used for the primary reaction and the same reaction conditions as the primary reaction. Some 40 μl of the product was electrophoresed and stained as before. Molecular size markers are shown on the left.

They then identified group-specific sequences for design of group-specific radiolabelled oligonucleotide hybridization probes. These probes could then be used specifically to detect 65K gene PCR product from each of the groups: *M. tuberculosis*/*M. bovis*, *M. avium*/*M. paratuberculosis* or *M. fortuitum* by Southern blotting or dot-blotting. The technique was capable of detecting as few as three mycobacterial cells mixed into 10^6 human mononuclear cells. Brissons-Noel *et al.* (1989) used this procedure to examine clinical specimens (sputa, gastric aspirates, abscess aspirates, tissue biopsy and blood) from TB and AIDS. All 13 culture positive TB specimens were also PCR positive for TB giving a sensitivity of 100%. However, two culture negative TB specimens (both gastric aspirates) were positive by PCR. These specimens were both from a single patient with recently diagnosed tuberculosis, but who had been receiving chemotherapy for four weeks. It seems likely therefore that in these samples the PCR was either amplifying DNA from very low numbers of viable live bacilli (not detected by culture), or was amplifying DNA from dead organisms. *M. tuberculosis* DNA was also detected in two of four AIDS blood samples from patients with mycobacterial infections (diagnosed by culture or presence of numerous acid-fast bacilli in stool).

Hartskeerl *et al.* (1989) utilized sequences encoding the 36 kD antigen of *M. leprae* to design a *M. leprae*-specific PCR reaction. The sensitivity of the PCR was estimated with dilutions of purified leprosy bacilli or *M. leprae* DNA. Dilutions of bacteria containing an estimated 2–10 cells, or DNA corresponding to 20 genomes (100 fg DNA) produced a visible PCR product on electrophoresis by ethidium bromide staining. *M. leprae* DNA could also be detected directly in liver homogenates of experimentally-infected armadillos.

Vary *et al.* (1990) used DNA sequence from the element IS*900* present in multiple copy in the genome of *M. paratuberculosis*, but absent in closely related strains of *M. avium* (Green *et al.*, 1989; McFadden *et al.*, 1990), to develop a *M. paratuberculosis*-specific PCR reaction for the diagnosis of Johne's disease. PCR product was detected with a radiolabelled IS*900*-specific hybridization probe by dot-blotting. Detection of 100 bacilli/ml of faeces was demonstrated. All culture-positive faecal samples from cattle with Johne's disease were also positive by PCR, indicating a 100% sensitivity. However a number of culture-negative faecal samples were detected as positive by PCR, which the authors suggest is due to a significant rate of false negatives by standard culture. Given the extremely high sensitivity of PCR, this may well be a problem in the initial evaluation of the PCR for clinical diagnosis. A commercial test system for rapid diagnosis of Johne's disease, based on this PCR assay, but using a non-isotopic labelling system, is shortly to be available in the USA and Europe from Idexx Corp. (Portland, Maine, USA).

The PCR is clearly capable of providing detection and identification of mycobacteria in clinical samples in a matter of a few hours from receipt of sample. If required, PCR reactions could be designed to give specific detection of any *Mycobacterium* — including differentiation of *M. tuberculosis* and *M. bovis* — all that is required is that DNA sequence differences (detected by RFLP work, above), be incorporated into the primer design. It should be remembered however that in common with most direct applications of DNA probes, present PCR systems may detect dead bacteria as well as live organisms. The technique may not therefore be suitable for applications where the presence of dead bacilli may be important, such as monitoring drug therapy. It is possible that DNA probes could be developed, targeted against RNA rather than DNA, that may be capable of detecting only live bacilli, since RNA is rapidly degraded in dead cells. Alternatively, quantitative PCR could be used to monitor DNA replication or messenger RNA induction.

The extremely high sensitivity of the PCR means that great care must be taken in sample preparation to avoid false positives due to contamination, particularly with PCR product. However, microbiologists are well used to working with the possibility of contamination in mind and the sterile procedures that are required in microbiological work can be modified to handle the PCR. To avoid PCR product contamination, we have a laboratory for sample and reagent preparation, that is located well away from the PCR laboratory. Reagents, equipment and even lab coats are never allowed to pass between the two laboratories. Small amounts of contaminating DNA present in reagents may be rendered unamplifiable by exposure of liquids for five minutes on a UV transilluminator (however if this is performed the solutions should not be used directly as free radicals formed may inactivate enzyme). Reagent blank controls are always run in parallel with PCR reactions; and when extracting DNA from samples for PCR, we normally perform a parallel DNA extraction of control samples.

2.3.4 Future prospects

DNA probes have the potential for reducing or even eliminating the need for lengthy culture in the mycobacterial laboratory. However, the information obtained from use of a DNA probe may not always be sufficient for clinical and epidemiological purposes and therefore culture may still need to be performed. No DNA probe presently available is specifically designed to predict mycobacterial drug sensitivity, although as mentioned above, species identification will often be sufficient to predict drug sensitivity. In many bacteria drug resistance is often due to acquisition of a specific drug resistance gene (e.g. β-lactamases in the Enterobacteriaceae), therefore DNA probes may be developed to detect the gene and therefore drug-resistant strains. However, in the mycobacteria, few genes encoding drug resistance determinants have been isolated (see Martin *et al.*, this volume). In addition, it is likely that in many cases, particularly of secondary or acquired drug resistance, the genetic changes associated with acquisition of drug resistance may be minor mutations, such as base substitutions that may affect drug sensitivity indirectly by, for example, modifying cell-wall permeability. Detection of these mutations may be difficult, unless they are found to be strain-specific when they may be detected indirectly by genetically linked RFLPs. DNA from dead bacteria may also be a problem in monitoring treatment efficacy. Therefore, development of DNA probes capable of measuring mycobacterial viability would be of considerable value; not only for mycobacteria such as *M. tuberculosis*, but particularly for *M. leprae* for which culture is not available.

The cost of DNA probes compared to traditional culture remains an obstacle to their introduction in many laboratories, particularly in the developing world where mycobacterial disease is most rife. However as non-isotopic probes become available, protocols become simpler, and large-scale production of probes, enzymes and other reagents reduces manufacturing costs, the cost of DNA probes should hopefully be reduced.

Finally, it is hoped that DNA probes will shortly provide answers to questions of mycobacterial involvement in diseases such as Crohn's disease, sarcoidosis and rheumatoid arthritis, where the role of mycobacteria has long been suspected, but never established. The sensitivity of DNA probes for detection of mycobacterial DNA in tissue, particularly when combined with the PCR should finally establish whether or not mycobacteria play a role in the aetiology of these diseases.

ACKNOWLEDGEMENTS

I would like to thank Dilip Banerjee for his advice on mycobacterial disease diagnosis. I would also thank Russel Enns of Gen-Probe for information concerning the Gen-Probe tests.

REFERENCES

Athwal, R. S., Deo, S. S. and Imaeda, T. (1984). *Int. J. Syst. Bacteriol.* **34**, 371–5.
Baess, I. (1979). *Acta Path. Microbiol. Scand. Sect. B.* **87**, 221–6.
Baess, I. (1983). *Acta Path. Microbiol. Scand. Sect. B.* **91**, 201–3.
Baess, I. (1984). *Acta Path. Microbiol. Scand. Sect. B.* **92**, 209–11.
Baess, I. and Weis Bentzon, M. (1978). *Acta Path. Microbiol. Scand. Sect. B.* **86**, 71–6.

Brissons-Noel, A., Gicquel, B., Lecossier, D., Levy-Fraubault, V., Nassif, X. and Hance, A. (1989). *Lancet* i, 1069–71.

Butcher, P. D., McFadden, J. J. and Hermon-Taylor, J. (1988). *Gut* 29, 1222–8.

Clark-Curtiss, J. E. and Docherty, M. A. (1989). *J. Infect. Dis.* 159, 7–15.

Collins, D. and De Lisle, G. (1984). *J. Gen. Microbiol.* 130, 1019–21.

Collins, D. and de Lisle, G. (1986). *Am. J. Vet. Res.* 47, 2226–9.

Cooper, G., Grange, J., McGregor, J. and McFadden, J. (1989). *Lett. Appl. Microbiol.* 8, 127–30.

Daniel, T. M. (1989). *Rev. Infect. Dis.* 11, (Suppl 2,) S471–8.

Drake, T. A., Hindler, J. A., Berlin, G. and Bruckner, D. (1987). *J. Clin. Micro.* 25, 1442–5.

Eisenach, K. D., Crawford, J. T. and Bates, J. H. (1986). *Am. Rev. Respir. Dis.* 133, 1065–8.

Eisenach, K. D., Crawford, J. T. and Bates, J. H. (1988). *J. Clin. Microbiol.* 26, 2240–5.

Ellner, P. D., Kiehn, T. E., Cammarata, R. and Hosmer, M. (1988). *J. Clin. Microbiol.* 26, 1349–52.

Gonzalez, R. and Hanna, B. A. (1987). *Diagn. Microbiol. Infect. Dis.* 8, 69–77.

Gross, W. M. and Wayne, L. (1970). *J. Bacteriol.* 104, 630–4.

Green, E., Tizard, M., Moss, M., Thompson, J., Winterborne, D., McFadden, J. and Hermon-Taylor, J. (1989). *Nucl. Acids Res.* 17, 9063–73.

Grosskinsky, C. M., Jacobs, W. R., Clark-Curtiss, J. E. and Bloom, B. (1989). *Infect. Immun.* 57, 1535–41.

Hampson, S., Portaels F., Thompson, J., Green E. P., Hermon-Taylor, J. and McFadden J. (1989). *Lancet* i, 65–8.

Hance, A. J., Grandchamp, B., Levy-Frebault, V., Lecossier, D., Rauzier, J., Bocart, D. and Gicquel, B. (1989). *Molec. Microbiol.* 3, 843–9.

Hartskeerl, R., Madelaine, Y., de Wit and Klatser, P. (1989). *J. Gen. Microbiol.* 155, 2357–64.

Heifets, L. (1989) [letter]. *J. Clin. Microbiol.* 27, 229.

Hurley, S. S., Splitter, G. and Welch, R. (1988). *Int. J. Syst. Bacteriol.* 38, 143–6.

Hurley, S. S., Splitter, G. A. and Welch, R. A. (1989). *J. Clin. Microbiol.* 27, 1582–7.

Kiehn, T. E. and Edwards F. F. (1987). *J. Clin. Microbiol.* 25, 1551–2.

Levy-Fraubault, V., Thorel, M-F., Varnerot, A. and Gicquel, B. (1989). *J. Clin. Microbiol.* 27, 2823–6.

McFadden, J. J., Butcher, P. D., Chiodini, R. and Hermon-Taylor, J. (1987a). *J. Gen. Microbiol.* 133, 211–14.

McFadden, J. J., Butcher, P. D., Chiodini, J. and Hermon-Taylor, J. (1987b). *J. Clin. Microbiol.* 25, 769–801.

McFadden, J. J., Butcher, P. D., Thompson, J., Chiodini, R. and Hermon-Taylor, J. (1987c). *Molecular Microbiology* 1, 283–91.

McFadden, J. J., Thompson, J., Hull, E., Hampson, S., Stanford, J. and Hermon-Taylor, J. (1988). In *Inflammatory Bowel Disease: Current Status and Future Approach* (ed. R. P. Macdermott), Elsevier Science Publishers B. V., Amsterdam, The Netherlands.

McFadden, J. J., Green, E., Moss, M., Tizard, M., Portaels, F. and Hermon-Taylor, J. (1990). Submitted.

Musial, C. E., Tice, L. S., Stockman, L. and Roberts, G. D. (1988). *J. Clin. Microbiol.* 26, 2120–3.

Pao, C. C., Lin, S. S., Wu, S. Y., Juang, W. M., Chang, C. H. and Lin, J. Y. (1988). *Tubercle* 69, 27–36.

Patel, R., Kvach, J. and Mounts, P. (1986). *J. Gen. Microbiol.* 132, 541–51.

Picken, R. N., Tsang, A. Y. and Yang, H. L. (1988). *Mol. Cell Probes* 2, 289–304.

Picken, R. N., Plotch, S. J., Wang, Z., Lin, B. C., Donegan, J. J. and Yang, H. L. (1988). *Mol. Cell. Probes* 2, 111–24.

Reddi, P. P., Talwar, G. P. and Khandekar, P. S. (1988). *Int. J. Lepr.* **56**, 592–8.

Roberts, G., Goodman, N., Heifets, L., Larsh, H., Lindner, T., McClatchy, K., McGinnis, M., Siddiqi, S. and Wright, P. (1983). *J. Clin. Microbiol.* **18**, 689–96.

Roberts, M. C., McMillan, C. and Coyle, M. B. (1987). *J. Clin. Microbiol.* **25**, 1239–43.

Saiki, R. K., Scharf, S., Faloona, F., Mullis, K., Horn, G., Erlich, H. A. and Arnheim, N. (1985). *Science* **230**, 1350–4.

Saiki, R. K., Gelfand, G. H., Stoffel, S., Scharf, S., Higuchi, S. J., Mullis, K., Horn, G. and Erlich, H. A. (1988). *Science* **239**, 487–91.

Saito, H., Tomioka, H., Sato, K., Tasaka, H., Tsukamura, M., Kuze, K. and Asano, K. (1989). *J. Clin. Microbiol.* **27**, 994–7.

Saxegaard, F. and Baess, I. (1988). *Acta Path. Microbiol. Scand. Sect. B.* **96**, 37–42.

Sherman, I., Harrington, N., Rothrock, A. and George, H. (1989). *J. Clin. Microbiol.* **27**, 241–4.

Shoemaker, S. A., Fisher, J. H. and Scoggin, C. H. (1985). *Am. Rev. Respir. Dis.* **131**, 760–3.

Shoemaker, S. A., Fisher, J. H., Jones, W. D. Jr and Scoggin, C. H. (1986). *Am. Rev. Respir. Dis.* **134**, 210–3.

Thierry, D., Cave, M., Crawford, J., Bates, J., Gicquel, B. and Guesdon, J. (1990). *Nucl. Acids Res.* **18**, 188.

Upholt, W. B. (1977). *Nucl. Acids Res.* **4**, 1257–65.

Vary, C., Andersen, P., Green, E., Hermon-Taylor, J. and McFadden, J. J. (1990). Submitted to *J. Clin. Microbiol.*

Wayne, L. G. (1978). *Ann. Microbiol. (Inst. Pasteur)* **129**, 13–27.

Whipple, D., Le Febvre, R., Andrews, R. and Thierman, A. (1987). *J. Clin. Microbiol.* **25**, 1511–15.

Woose, C. R. (1987). *Microbiol. Rev.* **51**, 221–71.

Yaeger, H., Smith, L. and Lemaister, C. (1966). *Am. Rev. Respir. Dis.* **95**, 998–1004.

Yoshimura, H. H. and Graham, D. Y. (1988). *J. Clin. Microbiol.* **26**, 1309–12.

Zainuddin, Z. and Dale, J. (1989). *J. Gen. Microbiol.* **135**, 2347–55.

8 Mycobacterial gene expression and regulation

Jeremy W. Dale and Abhay Patki

Department of Microbiology, University of Surrey, Guildford, Surrey GU2 5XH, UK

1 INTRODUCTION

Initial attempts to clone and express mycobacterial genes in *Escherichia coli* were met with a considerable degree of scepticism. This was partly due to the belief that organisms such as mycobacteria with a high G+C content in their DNA must use radically different transcriptional control systems from those in *E. coli*, and that this effect would probably extend to translational signals as well; it would therefore be impossible to get decent levels of expression in *E. coli*. This view was to some extent substantiated, for example, by the failure of mycobacterial cosmid libraries to complement any of a range of auxotrophic defects in an *E. coli* host strain (Clark-Curtiss *et al.*, 1985; Jacobs *et al.*, 1986b), although complementation of a citrate synthase mutation was demonstrated using an expression vector (Jacobs *et al.*, 1986a).

However, the defect in expression of mycobacterial genes in *E. coli* is far from absolute. Thole *et al.* (1985), working with a gene library of BCG in the lambda vector EMBL3, found that expression of selected antigens was weak but detectable by Western blotting. Similarly, Moss (1987), having constructed a BCG library in pBR322 and screened it with rabbit anti-BCG antiserum, was able to isolate clones that expressed the BCG 65 kD antigen, presumably using the mycobacterial transcription and translation signals. Therefore, at least some mycobacterial genes can be expressed in *E. coli*, although perhaps less efficiently than in their natural host; at the same time,

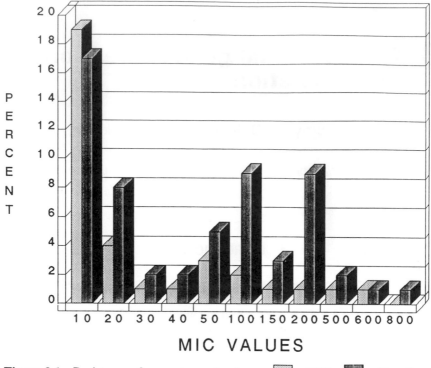

Figure 8.1   Resistance of promoter probe clones. ▫ , BCG; ▓ , *E. coli.*

there is considerable interest in those genes that do not appear to be expressed from typical consensus prokaryotic promoters.

Nevertheless, the concept of mycobacterial control systems as a block in the expression of these genes in *E. coli* led to the investigation of possible alternative cloning hosts, and attention was primarily directed towards *Streptomyces*, on the grounds that the DNA of these bacteria also has a high G+C content and that *Streptomyces* species appear to be less rigorous than *E. coli* in their promoter specificity. Furthermore, cloning systems in *Streptomyces* were reasonably well developed.

A comparative study was therefore carried out, using promoter probe vectors in both *E. coli* and *S. lividans* (in collaboration with workers at the John Innes Institute). The results of this study (Kieser *et al.*, 1986) have been widely cited as establishing that *Streptomyces* is to be preferred to *E. coli* as a host organism for expressing mycobacterial genes. The study also showed however that a considerable proportion of mycobacterial DNA fragments were in addition capable of acting as promoters in *E. coli*; about 20% of such

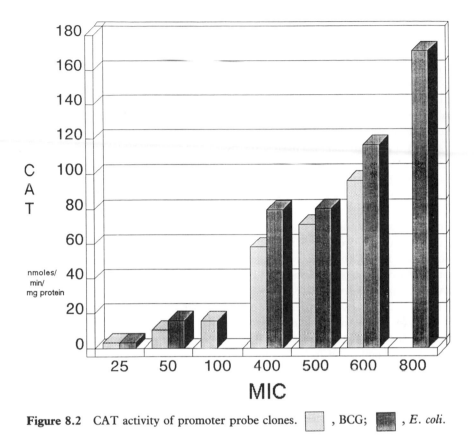

Figure 8.2   CAT activity of promoter probe clones. ☐ , BCG; ▨ , *E. coli.*

clones showed chloramphenicol resistance as compared to about 40% when *E. coli* DNA fragments were cloned in the same vector. In order to assess the relative strength of the mycobacterial and *E. coli* sequences as promoters, the resistance levels were tested for 170 clones of BCG DNA and 115 clones containing *E. coli* DNA. Figure 8.1 shows that while the BCG promoters were on the whole less active than the *E. coli* fragments, there were a minority of clones containing BCG DNA that were highly resistant to chloramphenicol, being capable of growth in chloramphenicol concentrations as high as 600 μg/ml. To confirm that the chloramphenicol resistance levels were indeed due to stimulation of transcription of the chloramphenicol acetyltransferase (CAT) gene, the specific activity of CAT was determined for cell extracts of six clones from each DNA source, with different resistance levels. Figure 8.2 confirms that there was a straightforward relationship between CAT levels and chloramphenicol resistance.

This has since been confirmed by the fact that a significant minority of mycobacterial genes have been shown to be expressed in *E. coli* apparently from their own promoters, most notably the 64–65 kD antigen of *Mycobacterium tuberculosis* (Shinnick, 1987), BCG (Thole *et al.*, 1987) and *M. leprae* (Mehra *et al.*, 1986), but also including other proteins such as the biotin carrier protein of several species (Collins *et al.*, 1987) and the *M. tuberculosis* 38 kD antigen (Andersen *et al.*, 1988). This is confirmed by the analysis of published DNA sequences, which shows the presence of typical consensus prokaryotic promoter sites in other genes in addition to those above (see Figure 8.3 and further discussion below).

We must be cautious about such studies however. Promoter-probe studies do not prove that the fragment in question actually has promoter activity in its original context. Nor does the computer identification of a sequence that is related to a conventional consensus promoter site establish that the sequence concerned actually functions as a promoter, even in *E. coli*, and still less that it is the natural promoter of that gene in the original mycobacterium. It should be noted that the computer analyses indicate a number of other potential promoter sites, in addition to those shown in Figure 8.3; these sites have been ignored because of their location, e.g. within the coding sequence of the gene. We need further analysis of transcriptional start points, e.g. by S1 analysis, in both organisms, and ultimately by direct RNA polymerase binding studies. This information is now acquiring additional importance with the acquisition of techniques for the introduction of foreign genes into mycobacteria (Snapper *et al.*, 1988; Zainuddin *et al.*, 1989).

The nature of promoter sequences has received more interest than other control mechanisms, largely due to the ease of computer analysis of DNA sequences. However, this is only one part of the process that intervenes between the DNA and the final active protein product. We need in particular to consider also the initiation of translation (i.e. the nature of the ribosome binding site and the initiation codon), the effect of codon usage on translation efficiency, and the likely effect of post-translational events on protein function. Of even more interest, potentially, are the genetic switches that control the activities of these systems; apart from the heat-shock proteins, very little is known at the molecular level of how the level of expression of specific mycobacterial genes may be modified by external conditions. It can be predicted that this will receive an increasing amount of attention over the next few years. This will happen both in the context of the genetic manipulation of mycobacteria and also in relation to understanding those processes that influence the behaviour of mycobacteria when growing *in vivo*, especially in the intracellular environment, and the role of these systems in the immunogenicity and pathogenicity of mycobacteria.

```
                        -35                                    -10                                          SD            start

consensus       TTGACA<----17-19------->TATAAT                                                      AGGAGGT<--4-7---->ATG

M. tuberculosis/M. bovis/BCG:

65K      TAGCGGGGTTGCCGTCACCCGTGACCCCG TTTCATCCCGA              --          TCCGGAGGAATCACTTCGCAATGGCC   ATGGCC
dnaJ                                                                        AGGCGAGAGGGGGTGACGCGAC       ATGGCC
19-22K                                                                      CAGGCAAGGAGGACAGG           GTGAAG
10-12K   GGGCGCCCTTGAGTGTCAGCACTTCATGTA TAGAGTGCTAGATG      - 145 bp -      CAAATAGTGGAGGGCTCCAATC       GTGGCG
TB38K    GGACTGTCGGGGACGTCAAGGACGCCAAGCGCGAAATTGAAGAGCA          -          CAGAAAGGTATGGC               GTGAAA
TB32K    GATGCGTTGAGATGAGGATGAGGGAAGC AAGAATGCAGCTTG        - (-25 bp) -     ATGAGGGAAGCAAGA              ATGCAG
BCGalpha CCGAATCGACATTTGGCCTCCACACACGTATGTTCTGGCCCG        - 15 bp -        ACAAGGGGCACAGGT              ATGACA
MPB70    GAAACACTTGAGGTGCGGCCCAGCAAGGGCTACAGGTTTTTCC       - 95 bp -        CGAAGGAGTGAACGGC             ATGAAG
MPB64    TCAGGCATCGTCGTCAGCAGCGCGATGCCC TATGTTTGTCGTCG     - 55 bp -        ACTCCCGGAGGAATTTCGAC         GTGCGC

M. leprae

65K      GAATTGCACTCGCCTTACGGGAGTGC TAAAAATGATCCTGGCACTCGCGATCGCGAGTGCC - 110 bp -
                                                             TCCTAATCCGGAGGAATCACTTCGCAATGGC
ML18K    GAAAACTTGTCTATCACAACTTGCATCAA TATATCGACCAGTG      - 60 bp -        GAGTGCGAGGTGACCACAC          ATGCTG
SOD      GGTGGGGCGGATCATNGCGCAGCGTTGATTATGCTAGTCG         - 45 bp -        AATCAAAGGAAGGAACGTC          GTGGCT
IRG      GATTCAATATAACCACTCTGGTCACACTAACCATACTCG TACCATCAACCGTG - 65 bp -   TTGAGGAGACTTCC               GTGCCG
ML36K    GTTGGGTTTCCTCTCCGGAGGGCGCACCGC TACGTTAGCGGGATG    - 125 bp -       CGCAGTGGAAGGTTACCC           ATGACC
```

Figure 8.3 Potential transcriptional and translational signals in mycobacterial genes. On the left hand side of the diagram, the potential transcriptional signals (-35 and -10 regions) as identified by the procedure of Staden (1984) are shown in bold. Other potential regulatory sequences discussed in the text are shown underlined, namely 'heat-shock promoters' (65K, 10–12K), 'phosphate box' (TB38K — see also Figure 8.4), and 'iron-regulatory sequence' (M. leprae IRG). On the right hand side, the DNA sequences corresponding to the putative translational signals (Shine–Dalgarno sequences and associated translational start codons) are shown in bold. An alternative start codon (underlined) has also been suggested for the M. leprae 65K gene. The E. coli consensus for each of these signals is also shown for comparison. For the identity of these genes, and references, see Table 8.1.

2 TRANSCRIPTIONAL CONTROLS

2.1 Promoter specificity

The specific recognition of promoter sites by RNA polymerase, largely mediated by the associated sigma factor, plays an important role in the control of DNA transcription. This ensures first of all that transcription only occurs from defined sites and in a specific direction, and secondly, the nature of the promoter will influence the affinity of the RNA polymerase for that site, and hence determine the efficiency of transcription which is ultimately the major determinant of the level of gene expression. It is convenient to consider promoter sites in terms of the two most conserved regions, at -35 and -10 with respect to the start point of transcription. In *E. coli*, and many other bacteria, most of the genes that are expressed during normal vegetative growth have recognizably similar sequences at these positions (TTGACA and TATAAT, respectively), usually separated by 17–19 bases. This combination is referred to as the typical consensus prokaryotic promoter, or more precisely, the promoter that is recognized by RNA polymerase when the enzyme is combined with one specific sigma factor, sigma 70. Under certain circumstances, sigma 70 is replaced by other sigma factors, and the promoter specificity of the RNA polymerase is altered so that a different group of genes is expressed (see below).

Although most reports concentrate on the identification of sequences related to the -10 and -35 regions, the RNA polymerase makes contact with the DNA over a much longer region, even if the requirements for such contact are less strict. This means that in addition to these recognizably conserved sequences, the nature of the DNA sequence over a region of about 60 basepairs (-50 to $+10$) is likely to have some influence over the affinity of the RNA polymerase for the promoter site, and its ability to initiate transcription from this position.

In order to locate possible promoter sites, we have used the procedure of Staden (1984) to screen all the available mycobacterial gene sequences, as listed in Table 8.1; the results of this analysis are shown in Figure 8.3. With a few exceptions, it was possible to identify a putative promoter site for each gene; some of these sites correspond to those identified by the authors of the original paper, while others are different. However, there is a considerable degree of flexibility in the 'rules' for determining a potential promoter site, and there is little constraint on its position except that it must be upstream of the gene in question and in the correct orientation. However, most of the sites listed are acceptably close to the presumed translational origin.

Table 8.1 Sequenced Mycobacterial Genes.

Databank code	Identity	Mnemonic	Reference
M. tuberculosis:			
MSGTCWPA	65 kD antigen (hsp60, *groEL*)	TB 65K	Shinnick, 1987
MTDNAJ	*dnaJ* analogue	TB dnaJ	Lathigra *et al.*, 1988
MSGANT19	19–22 kD antigen	TB 19–22K	Ashbridge *et al.*, 1989
MSG10KAG	10–12 kD antigen (*groES*)	TB 10–12K	Baird *et al.*, 1989
MTBCGA	10–12 kD antigen (*groES*)	TB 10–12K	Shinnick *et al.*, 1989
M30046	38 kD antigen (*pstS* analogue)	TB 38K	Andersen and Hansen, 1989
M27016	32 kD antigen	TB 32K	Borremans *et al.*, 1989
M. bovis:			
X15803	19–22 kD antigen	MB 19–22K	Collins *et al.*, 1990
BCG:			
MSGBCG	64 kD antigen (hsp60, *groEL*)	BCG 64K	Thole *et al.*, 1987
MSGBCGA	alpha antigen (30/31 kD)	BCG alpha	Matsuo *et al.*, 1988
MBMPB57	10–12 kD antigen (MPB57)	BCG 10–12K	Yamaguchi *et al.*, 1988
–	18 kD antigen	BCG MPB70	Terasaka *et al.*, 1989
–	23 kD antigen	BCG MPB64	Yamaguchi *et al.*, 1989
M. leprae			
MSGANTM	65 kD antigen (hsp60, *groEL*)	ML65k	Mehra *et al.*, 1986
MSGANT18K	18 kD antigen	ML18K	Booth *et al.*, 1988
MLEPSOD	28 kD antigen (a) (superoxide dismutase)	SOD	Thangaraj *et al.*, 1989
MSG28KDAG	28 kD antigen (b) (hypothetical iron transport protein)	IRG	Cherayil and Young, 1988
–	36 kD antigen	ML36K	Thole *et al.*, 1990

The list contains only those genes where complete published sequences are available, and where a protein product is known, or there is substantial evidence for such a product. Open Reading Frames with hypothetical products have not been included in the analyses in this chapter. Codes starting MSG are GenBank locus numbers; those starting MT, MB, ML are from EMBL. Other codes are accession numbers.

Two genes are shown without any identified transcriptional signal: the *dnaJ* and 19–22 kD genes. Of these, the former is an open reading frame downstream from the 71 kD (*dnaK*) gene, and identified as *dnaJ* by its homology with *dnaJ* of *E. coli*. In *E. coli*, these two heat-shock genes are located on a single operon, and this may well be true of *M. tuberculosis* as well, which would account for the apparent absence of a promoter. The 19–22 kD gene of *M. tuberculosis* (and *M. bovis*) appears not to be preceded by a consensus promoter sequence; it is not known how this gene is transcribed in mycobacteria, but it may also be a stress protein and hence may have a different type of promoter (see Young *et al.*, this volume). It does not appear to be expressed unaided in *E. coli*.

The location of expression signals associated with the 32 kD antigen gene is also uncertain. One possible consensus promoter site was detected. However, RNA polymerase using this site would start transcription from a point beyond the most likely translational start point. There are several alternative start codons in the correct reading frame, but the one indicated (ATG$_{234}$ in the numbering of Borremans *et al.*, 1989) is the only one associated with a good Shine and Dalgarno sequence. Translation initiating from this point would give rise to a rather unorthodox signal peptide of 42 amino acids. Another possible start codon (ATG$_{273}$) would result in a more typical signal peptide of 29 amino acids, but there is no consensus ribosome binding site. Borremans *et al.* (1989) did not obtain expression in *E. coli* of the mature protein from recombinants carrying the upstream region, indicating that either transcriptional or translational signals are not functional in *E. coli*, and presumably that alternative expression signals are used by *M. tuberculosis*. This is discussed in more detail by Borremans *et al.* (1989).

The *M. tuberculosis* 65 kD and 10–12 kD antigens are major heat-shock proteins, analogous to GroEL (hsp60) and GroES respectively, and are preceded by potential heat-shock promoter sequences for recognition by the alternative sigma factor, sigma 32 (Cowing *et al.*, 1985). These sequences are underlined in Figure 8.3, and it should be noted that they overlap with the typical (sigma 70 recognized) promoter. It is not unusual for genes to have more than one class of promoter, and for the sequences to overlap, when transcription is required both in normal circumstances and (in this case) after heat shock. A more detailed account of stress proteins, their function and regulation, is provided by Young *et al.* (this volume).

The DNA sequences upstream from the 65 kD genes of *M. tuberculosis* and *M. leprae*, although showing a considerable degree of homology, are much more divergent than the coding sequences themselves. In the case of the *M. tuberculosis*/BCG sequence, two possible consensus (sigma 70) promoters have been identified: that shown in Figure 8.3 (see also Thole *et al.*, 1987), and one further upstream (Shinnick, 1987). The latter was not detected by our analysis, and is therefore not shown. The converse was true for the *M. leprae* 65 kD gene. Only the upstream site was identified as a potential promoter, while in the region corresponding to the other site, the differences in the

DNA sequence between the two genes make the *M. leprae* sequence less closely related to either the sigma 70 or sigma 32 (heat-shock) promoters.

The transcriptional control signals of the two remaining genes (*M. tuberculosis* 38 kD and *M. leprae* 28 kD(b)) are discussed in separate sections below.

Although we do not yet have direct evidence of the real nature of the promoters and transcriptional start sites for any of these mycobacterial genes, the identification of so many putative promoters and regulatory sequences related to corresponding sequences in *E. coli* should give us confidence that the basic mechanisms involved in mycobacteria are unlikely to be dissimilar to those in other bacteria. This also indicates that it should be possible to search directly for certain genes by exploiting the likely regulatory mechanisms associated with such genes.

2.2   *M. tuberculosis* 38 kD antigen: a phosphate-regulated gene?

As with several other mycobacterial genes (see above) the *M. tuberculosis* 38 kD gene can be expressed in *E. coli* from a lambda gt11 recombinant, independently of IPTG addition (Andersen *et al.*, 1988). This indicates that transcription can be initiated from within the mycobacterial insert, presumably (but not conclusively) from the natural promoter of the gene. However, analysis of the sequence (Andersen and Hansen, 1989) does not reveal any regions upstream from the putative translational start position that resemble a consensus prokaryotic promoter.

Andersen and Hansen (1989) also showed that the deduced amino acid sequence of the 38 kD antigen has a potentially significant degree of homology to the *E. coli* phosphate-binding protein PstS (also known as PhoS) (Surin *et al.*, 1984). This is a periplasmic protein that forms part of the phosphate-repressible high-affinity phosphate specific transport (Pst) system of *E. coli*, in contrast to the constitutive low-affinity, inorganic phosphate transport system (Pit). See Young *et al.* (this volume) for further discussion of the role of phosphate-binding proteins.

The *pstS* gene is part of the *pst* operon (Surin *et al.*, 1985; Amemura *et al.*, 1985), which is in turn part of the *pho* regulon. This consists of a number of genes at different loci which are co-regulated in response to the presence of phosphate, so that the growth of the organism in the presence of limiting amounts of phosphate results in the simultaneous induction of many different proteins with the function of attempting to scavenge phosphate from any traces of phosphate-containing nutrients. These genes include *phoE* (coding for an outer membrane porin that facilitates the uptake of phosphate), *phoA* (periplasmic alkaline phosphatase), and *ugp*, which encodes a binding-protein dependent uptake system for glycerol-3-phosphate (Tommassen *et al.*, 1987).

The coordinated control of scattered genes can be achieved by means of related control elements in the promoter region of each gene (or operon), so that each promoter can be regulated by the same protein. In the case of the *pho* regulon, a comparison of the sequences of several of the genes involved (Surin *et al.*, 1984; Makino *et al.*, 1986; Tommassen *et al.*, 1987) has demonstrated the presence of an atypical promoter containing a standard Pribnow (-10) box but instead of the usual -35 region there is a different, 17 basepair, element termed a *pho* box.

Andersen and Hansen (1989) did not find any homologies to the *pho* box in the upstream sequence of the *M. tuberculosis* 38 kD antigen gene. We have re-examined the sequence, and find that there is in fact a region that bears a partial resemblance to a *pho* box (see Figure 8.4). There is an additional occurrence of the GTC repeat further upstream in this case, so that if the *pho* box consensus is represented as CTGTC (A/T)$_6$ CTGTC, then the structure of this region of the 38 kD gene is ACGTC (G/C)$_5$ ACTGTC (G)$_5$ ACGTC. If one concedes the switch of the central portion from the *E. coli* AT-rich situation to a GC-rich core, then it is conceivable that this sequence may represent the mycobacterial equivalent of the *pho* box. It will be of interest to see whether similar sequences are detected upstream from other phosphate-regulated genes.

```
A
                                 T       T T
    pho box consensus      CTGTCATAAAACTGTCA  -  11  -   TATAAT
                           *****      ** ****         * * *
    38kD gene              CTGTCGGGGGAC GTCA  -  15  -   GAAATT


B
    Outline pho box consensus:                    CTGTC  (A/T)6  CTGTC
    38kD gene pho box?:        ACGTC  (G/C)5  ACTGTC  (G)5   ACGTC
```

Figure 8.4 A possible *pho* box upstream from the *M. tuberculosis* 38 kD antigen gene. (A) Comparison of the putative *pho* box and -10 region of the 38 kD gene with the *E. coli* consensus sequences. (B) The *E. coli pho* box can be represented as a repeated CTGTC sequence separated by an AT-rich core. The putative mycobacterial homologue has three copies of a similar partial repeat separated by a G- or GC-rich core.

Tommassen *et al.* (1987) demonstrated that the *E. coli phoE* gene required an additional upstream element for efficient expression, and that this upstream element had homology to the *pho* box, but in the opposite orientation. We have been unable to find such a structure in this case.

2.3   Iron-regulated gene expression

Another group of regulatable genes with undoubted biological significance is that concerned with the uptake of iron. The acquisition of iron from the surrounding medium poses considerable problems for a bacterial cell. Ferrous iron, Fe(II), is readily soluble in water but is stable only in a strictly non-oxidizing environment. In the presence of air, it is oxidized to the ferric state, Fe(III), which is virtually insoluble (10^{-18}M). All organisms therefore, with the possible exception of strict anaerobes, require a mechanism for solubilizing extracellular Fe(III). Pathogenic bacteria face an additional problem in that they have to compete with their host for this supply of iron (Finkelstein *et al.*, 1983; Bagg and Neilands, 1987). In the human body, three quarters of the iron is found in haemoglobin and most of the rest is complexed with ferritin in the liver. The iron-carrying compound that is responsible for supplying cells with iron is the glycoprotein transferrin, which is normally about one-third saturated with Fe(III).

Bacteria acquire Fe(III) from their environment by means of chelating agents known as siderophores (some of which are products of other micro-organisms), combined with a mechanism for transporting the chelated Fe(III) into the cell. Mycobacteria produce two types of iron-chelating compound — the extracellular exochelins and the cell-bound mycobactins (Ratledge, 1982, 1984; Barclay and Ratledge, 1988). In addition, the presence of citrate in the culture medium can enable bacteria to take up iron in the form of ferric citrate. In *E. coli*, this process is mediated by membrane proteins coded for by the *fec* genes. Citrate-mediated iron uptake has also been demonstrated in mycobacteria (Messenger and Ratledge, 1982). It is not the purpose of this chapter to discuss the details of the mechanisms of iron uptake, but to consider the role of iron in regulating these processes and other aspects of gene expression.

Regulation of the uptake of iron is especially important to the cell since surplus iron may interact catalytically with superoxide and peroxide to produce hydroxyl radicals, with extremely damaging consequences (Bagg and Neilands, 1987). It has been known for some time that the production of siderophores is regulated in response to the iron content of the medium (see Neilands, 1981; Bagg and Neilands, 1987), i.e. siderophore production is 'repressed' by the presence of iron. This applies to the production of mycobactins and exochelins by mycobacteria (Ratledge, 1984) as well as to better-characterized systems in other organisms, notably aerobactin production.

Envelope proteins are usually required for the transport of siderophores, and some of these are also known to be iron-regulated. The finding that the production of some envelope proteins of *M. smegmatis* is iron-regulated (Hall *et al.*, 1987) may indicate that these are also involved in iron transport, and this is supported by the finding that antibodies to one of these components inhibited iron uptake by *M. smegmatis* (Hall *et al.*, 1987; Barclay and Wheeler, 1989). The regulation of these systems in bacteria other than mycobacteria is now reasonably well understood, and will be considered later in this chapter.

In addition to the iron uptake mechanism, other genes in a variety of organisms are known to be expressed selectively under iron-limiting conditions. The best known of these is the diphtheria toxin, produced by strains of *Corynebacterium diphtheriae* that are lysogenic for certain bacteriophages. The similarities that exist in other respects between corynebacteria and mycobacteria provide a justification for including a discussion of this system, in the hope that we may be able to draw some inferences about possible mechanisms in mycobacteria, as yet uncharacterized.

Over a decade ago, Murphy *et al.* (1976) suggested that production of diphtheria toxin was subject to negative regulation by a repressor protein, using iron as a co-repressor. This hypothesis was supported by the isolation of presumed operator-constitutive mutant phages in which toxin production was not sensitive to the presence of iron. Later (Kanei *et al.*, 1977), the isolation of chromosomal mutants that were not iron-repressible not only provided further support for this hypothesis but also indicated that the putative repressor gene was located on the bacterial chromosome and did not form part of the phage genome. Cloning and sequencing the *tox* gene (Kaczorek *et al.*, 1983; Greenfield *et al.*, 1983; Ratti *et al.*, 1983) showed the presence of a consensus bacterial promoter site, and low levels of expression were indeed obtained in *E. coli*. (Another parallel with mycobacteria is that the initiation codon is GTG rather than ATG, see below). Leong and Murphy (1985) confirmed the location of the transcriptional start point by S1 mapping and also showed that in *C. diphtheriae* the specific transcript was only produced under iron-limiting conditions. More recently (Tai and Holmes, 1988), the regulatory activity of the promoter region was studied by using a *gal*K transcriptional fusion in *E. coli*; the production of galactokinase was enhanced by growth in iron-depleted medium, and they confirmed that regulation takes place at the transcriptional level. These results are significant in that they indicate that *E. coli* contains a regulatory system that is capable of mediating the repression of the *C. diphtheriae* toxin gene by iron.

In *E. coli*, the synthesis of iron uptake systems is regulated by the product of a gene known as *fur*, which stands for Ferric Uptake Regulation (Hantke, 1981, 1984). Genes controlled in this manner include the aerobactin operon of ColV plasmids (*iuc*), genes involved in the production and transport of enterobactin (*ent*, *fep*, *fes*), or the transport of exogenously provided siderophores (*fhu*) (see de Lorenzo *et al.*, 1987; Bagg and Neilands, 1987, for further references), as well as the Shiga-like toxin or 'Verotoxin' (*slt*; Calderwood and Mekalanos, 1987; de Grandis *et al.*, 1987). The *Pseudomonas aeruginosa* exotoxin A is similarly iron-regulated (Sokol *et al.*, 1982).

Homologies in the promoter region of these genes, and of the *fur* gene

		-35		-10
iucA P1	CATTTCTCATT	GATA	ATGAgAATCATTATt	GACA
sltA	AGCCTCTCTTT	GAat	ATGATtATCATTtTC	ATTA
fhuA	TATTATCTTAT	ctTt	ATaATAATCATTcTC	GTTT
fepA	TATATTAGTAA	tATt	ATGATAActATTtgC	ATTT
fur	CGTGGCAATTC	tATA	ATGATAcgCATTATC	TCAA
fhuE	TGAATGCGTAT	atTt	cTcATttgCATTtaC	AAAC
tonB	TTATTGAATAT	GATt	gctATttgCATTtaa	ATCG
tox	TAATTAGGATA	GcTt	taccTAATtATTtTa	TAGC
		---------⟩ ⟨--- ----		
consensus		GATA	ATGATAATCATTATC	
M.leprae 28kD	CAATTACCTCAcGATtcAatATAAcCAcTcTg			GTCA
(IRG)				

Figure 8.5 Hypothetical Fur-binding sites of iron-regulated promoters. For *iucA* the sequence shown is that of the primary Fur binding site in the P1 promoter region as determined by footprinting (de Lorenzo *et al.*, 1987) with -10 and -35 regions. The other sequences are aligned on the basis of their similarity with the *iuc*A sequence and the proposed consensus (which also takes into account other sequences not shown here). Lower case letters indicate bases that do not correspond to the consensus sequence. The sequences shown vary in their position relative to the transcriptional start point. The region of dyad symmetry is shown for the consensus only; dyad symmetry is present in most of these sequences but in different positions. See Tai and Holmes (1988) for further details and references.
28 kD = *M. leprae* 28 kDa protein (Cherayil and Young, 1988); the region shown is -144 to -109.

itself, are sufficient to suggest a consensus sequence required for the binding of the Fur protein (Figure 8.5); this sequence contains a highly AT-rich palindrome. This was confirmed by footprinting studies (de Lorenzo *et al.*, 1987) which defined a binding site for the Fur protein in the promoter region of the aerobactin operon, and also showed that the binding of the Fur protein was considerably enhanced by the presence of divalent cations (Mn^{2+}). In addition, Calderwood and Mekalanos (1988) showed that insertion of a synthetic oligonucleotide, corresponding to the proposed Fur-binding consensus, into a site adjacent to a promoter could render the gene repressible by the presence of iron in the medium.

When the *tox-galK* fusions referred to above were introduced into a *fur* mutant strain of *E. coli*, galactokinase was produced at a high rate, and production was unaffected by the presence of iron. This indicates that the *C. diphtheriae tox* gene can be controlled by the *E. coli* Fur protein, and this is supported by examination of the sequence which shows that the promoter region of the diphtheria toxin gene shares a considerable degree of homology with the consensus sequence for the *E. coli* Fur-binding site (Tai and Holmes, 1988; see Figure 8.5). The putative repressor of the *C. diphtheriae* *tox* gene may therefore be similar to the *E. coli* Fur protein, implying a considerable degree of conservation in the mechanism for the iron regulation of genes in diverse species.

Is this information relevant to mycobacteria? It would seem to be very likely that the control mechanisms operating for the production of myco-bactins and exochelins, the iron-regulated envelope proteins, and possibly some as yet unidentified virulence factors, will prove to be similar at least in overall concept to those that exist in organisms as diverse as *E. coli* and *C. diphtheriae*, especially in view of the latter's taxonomic relationship to mycobacteria. However, one feature that *C. diphtheriae* does not share with mycobacteria is a high G+C DNA base composition: the *tox* region is only 43% G+C, which is considerably less than the 62–70% for mycobacterial DNA — or even the 51–59% overall composition for DNA from coryne-bacteria (Goodfellow and Minnikin, 1984). The proposed consensus Fur-binding site (Figure 8.5) is highly AT-rich and might therefore be expected not to occur in mycobacteria.

However, a computer search of known mycobacterial sequences revealed one region with a considerable degree of homology to the Fur-binding site. This is upstream from the gene coding for the 28 kD antigen of *M. leprae* that was identified and sequenced by Cherayil and Young, 1988 (see Figure 8.5). Although the match with the proposed consensus is far from complete (12/19 if the single base insertion is allowed), inspection of Figure 8.5 shows that several of the characterized iron-regulated genes have weaker relationships to the consensus. The potential significance of the match is enhanced by the location of this region of the 28 kD gene, since it overlaps with the -35 region of the potential promoter site (see Figure 8.3), and is thus situated in precisely the position expected for such a control element.

Although the 28 kD gene is not known to be iron-regulated in *M. leprae* (and indeed such studies are not easy for non-cultivable bacteria), this comparison indicates that it is likely to be repressed by the presence of iron. This carries the further implication that in *M. leprae* (and presumably other mycobacteria) iron regulation of gene expression is mediated by a protein homologous to the Fur protein of *E. coli*.

This observation is lent an additional degree of significance by the suggestion of Cherayil and Young (1988) that the protein coded for by this gene is associated with the cell envelope. This combination of hypotheses leads to the inference that this protein may be involved with the iron uptake mechanism of *M. leprae*. This is further supported by the observation that a 29 kD membrane protein is selectively expressed under low iron conditions by several mycobacteria, and that antisera to this protein inhibited iron uptake by *M. smegmatis* (Hall *et al.*, 1987; Barclay and Wheeler, 1989).

If it is confirmed that the 28 kD *M. leprae* antigen is an iron transport protein, this would be a finding of considerable significance, since this would be a key component of the pathogenicity of the bacterium. While it will be difficult to verify this directly in *M. leprae*, similar genes can now be sought in other mycobacteria and their function examined, together with experiments to test the effect of iron on the expression of this gene and the Fur-binding capacity of this region in *E. coli* and in cultivable mycobacteria. This may then lead to the development of a novel range of antibiotics that act by affecting the iron transport mechanism.

2.4   Inducible and repressible enzymes

The previous sections have shown that the *M. tuberculosis* 38 kD and *M. leprae* 28 kD(b) antigen genes are associated with DNA sequences that suggest the possibility of specific regulatory mechanisms, without such control having been demonstrated directly. The converse is true for a number of other mycobacterial enzymes: apparent induction and/or repression has been known for many years (see reviews by Ratledge, 1982; Segal and Edwards, 1984) without the molecular basis of these control mechanisms having been defined.

One example is the utilization of glycerol as a carbon source which occurs via glycerol-3-phosphate and dihydroxyacetone phosphate; the enzymes involved are glycerol kinase and glycerol 3-phosphate dehydrogenase, respectively. The latter enzyme is induced with glycerol, and to a lesser extent with its own substrate, glycerol 3-phosphate, but the production of glycerol kinase is constitutive. Chloramphenicol has the unusual effect of not only abolishing induction, but also of eliminating the substantial basal level of glycerol 3-phosphate dehydrogenase (Segal and Edwards, 1984). This is interpreted as indicating a high turnover rate for the enzyme, coupled with the presence of endogenous inducer.

Glycerol is not used as a growth substrate in the presence of other carbon sources such as glucose or glutamate. Such effects are often due to catabolite repression which arises from the reduction in the level of cyclic AMP (cAMP) in the presence of glucose; in such cases, cAMP (in combination with a receptor protein) stimulates transcription of the genes concerned. Although the level of cAMP in mycobacteria has been shown to drop markedly following the addition of glucose (see Ratledge, 1982), the glucose effect in this case seems to be a more indirect one (Bowles and Segal, 1965), while glutamate directly represses the production of glycerol 3-phosphate dehydrogenase.

The utilization of other carbon sources is also subject to induction, with the specificity of induction of a variety of pentose isomerases and permeases having received considerable attention. Of particular interest is the ability of D-galactose to act as a gratuitous inducer for the L-arabinose isomerase (Izumori et al., 1978), suggesting that the isomerase gene forms part of an operon with the L-arabinose permease gene; the latter enzyme is also induced by D-galactose, and is capable of transporting D-galactose into the cell (Ratledge, 1982).

The pathways of biosynthesis of amino acids are also subject to control mechanisms that appear to parallel those existing in other bacteria. The 'aspartate family' is usually subject to complex and interacting patterns of feedback inhibition/repression, necessitated by the branched nature of the pathway resulting in the production of several amino acids (lysine, methionine, threonine and isoleucine). General control mechanisms operate at the aspartokinase and homoserine dehydrogenase steps, supplemented by specific controls on the individual pathways distal to the branch points. Karasevich and Butenko (1966: cited by Segal and Edwards, 1984) showed that a mycobacterial aspartokinase was repressed by threonine and methionine while homoserine dehydrogenase was repressed by threonine and valine. Valine enters the picture as it shares the enzymes of the biosynthetic pathway responsible for the production of isoleucine. Feedback inhibition/repression of different stages by valine or isoleucine alone is expected therefore to result in that amino acid being inhibitory to growth; this is commonly found in many bacteria, and mycobacteria are no exception (Segal and Edwards, 1984).

More recently, the control of aspartokinase and homoserine dehydrogenase in M. smegmatis has been analysed in more detail (Sritharan et al., 1989). Each of the products of the pathway was shown to act as a feedback inhibitor and/or repressor of one or both of these enzymes. Three isoenzymes of aspartokinase were purified and shown to be differentially regulated by the end-products of the pathway with the general conclusion that lysine biosynthesis was primarily regulated at the aspartokinase level while threonine and methionine synthesis was controlled by regulating the homoserine dehydrogenase. This work was supplemented by a study (Sritharan et al., 1990) of M. avium grown axenically in comparison with the same organism grown in vivo and with M. leprae obtained from armadillo tissue. Mycobacteria are capable of assimilating the amino acids of this family from the supply available within the host tissue, and the results obtained indicate that this pathway (or specifically the key stages, aspartokinase and homoserine dehydrogenase) is repressed in the in vivo grown mycobacteria.

Another inducible system that has received attention is the production of amidases, notably by M. smegmatis. Various mycobacterial species have the ability to produce enzymes that are able to hydrolyse different ranges of amides and this has been used as the basis of classification and identification systems. However, the inducible nature of these enzymes can lead to considerable variability in the results obtained, which is a prime reason for the decline in popularity of this as the sole basis for identification. Draper (1967) showed that an amidase with preferential activity against formamide, and a lower level of activity against a range of other aliphatic amides, was induced 50–100-fold by growth in the presence of acetamide. This was extended by Iwainsky and Sehrt (1970) who demonstrated, in addition to a broad spectrum aliphatic amidase, two apparently separate amidases: a succinamidase that was specifically induced by succinamide, and a urease that was inducible by most amides tested, including, as the most effective inducers, asparagine and glutamine. The fact that the increased enzyme activity was prevented by the addition of chloramphenicol supports the notion that the phenomenon was a genuine induction. The separate identity of the urease was confirmed by a subsequent study (Iwainsky and Sehrt, 1971) showing that several amino acids induced an enzyme with specific activity against urea and little or no ability to hydrolyse aliphatic amides. Synthesis of this enzyme was repressed by the presence of ammonium ions.

The amidase group of enzymes represents one of the clearest examples of inducible enzymes in mycobacteria. In many other cases the induction ratio is so low as to cast doubts on the reality of the induction. This is certainly true of enzymes such as penicillinase (beta-lactamase); the presence of this enzyme has been demonstrated in many mycobacterial species (Kasik, 1979) and in several cases has been claimed to be inducible. A more recent example is the beta-lactamase of *M. fortuitum* (Nash et al., 1986) which appeared to be produced at an elevated level after growth in the presence of an inducing agent (ceftizoxime). However, the increase in activity was small (less than threefold), and may not represent true induction. The usual picture with beta-lactamases from other mycobacteria is of constitutive synthesis of this enzyme (Kasik, 1979; Choubey and Gopinathan, 1986). In contrast, the high induction ratio of the aliphatic amidase makes it eminently exploitable by molecular biologists. However, the biological significance of the amidases is unclear, and the regulatory mechanisms of all these inducible systems remain to be elucidated.

3 TRANSLATIONAL CONTROL AND POST-TRANSLATIONAL EFFECTS

In the natural situation, it is to be expected that translational controls will be less significant overall than transcriptional controls. However, there are certain features that are required for effective expression, or that may modulate the level of expression.

In bacteria, translation is normally initiated following the attachment of ribosomes to a specific ribosome-binding site on the mRNA. The principal recognizable feature of such a ribosome binding site is the presence of a purine (A+G) rich sequence of about seven bases, known as a Shine and Dalgarno sequence, followed after an interval of 4–7 bases, by a start codon, usually AUG. The binding of ribosomes to such a region is facilitated by the complementarity between the 3′-end of the 16s ribosomal RNA and the Shine and Dalgarno sequence on the mRNA, enabling a transient basepaired structure to form. As will be seen from Figure 8.3, there is little difficulty in finding suitable ribosome-binding sites for mycobacterial genes that have been sequenced, which provides comforting reassurance that in this respect as well mycobacteria conform to the usual rules. The 65 kD genes appear to have a longer than usual separation between the best Shine and Dalgarno sequence and the start codon, which would be expected to have an adverse effect on translation, and it could be that the alternative version, three bases downstream, (AGGAAUC) is preferred *in vivo*.

The circularity of the argument needs to be noted however. In most cases, there are several possible start codons in the same reading frame and the real start codon has not been identified by direct experiment. The start codons indicated in Figure 8.3 have therefore been chosen on the basis that they are adjacent to a recognizable Shine and Dalgarno sequence.

A further feature of the translational initiation signals that should be noted is that in the sequences so far reported, GUG is used as a start codon nearly as frequently as AUG. Although the use of GUG as a start codon has only been confirmed by comparison with N-terminal amino acid analysis for the 10–12 kD antigen (Yamaguchi *et al.*, 1888; Shinnick *et al.*, 1989; Baird *et al.*, 1989) the evidence for a GUG initiation codon is strong in the cases of the 19–22 kD gene (Ashbridge *et al.*, 1989; Collins *et al.*, in preparation) and the *M. leprae* 28 kD antigen (Cherayil and Young, 1988). While the number of sequences recorded is still too small to make any firm predictions, the use of GUG as a start codon would be consistent with the higher G+C content of mycobacterial DNA.

3.1   Codon usage

Most amino acids can be coded for by more than one codon, and the relative frequency with which these different synonymous codons occur is referred to as the codon usage. In an organism like *E. coli*, many genes seem to have very little preference for particular codons, i.e. there is little codon bias. However, highly expressed genes in *E. coli* show a higher degree of codon bias, which

Table 8.2 Codon Usage in *Mycobacterium tuberculosis* (including *M. bovis*, BCG).

a.a.	codon	number	tot	%	a.a.	codon	number	tot	%	a.a.	codon	number	tot	%
F	UUU	6		8	P	CCU	9		6	N	AAU	14		13
F	UUC	65	71	92	P	CCC	44		28	N	AAC	98	112	88
L	UUA	2		1	P	CCA	13		8	K	AAA	12		10
L	UUG	39		18	P	CCG	91	157	58	K	AAG	106	118	90
L	CUU	5		2	T	ACU	10		6	D	GAU	23		17
L	CUC	41		19	T	ACC	116		64	D	GAC	109	132	83
L	CUA	8		4	T	ACA	11		6					
L	CUG	124	219	57	T	ACG	44	181	24	E	GAA	24		19
I	AUU	20		19	A	GCU	29		9	E	GAG	104	128	81
I	AUC	82		77	A	GCC	153		49	C	UGU	7		25
I	AUA	4	106	4	A	GCA	27		9	C	UGC	20	27	74
M	AUG	43	43	100	A	GCG	104	313	33					
V	GUU	20		10	Y	UAU	10		14	W	UGG	31	31	100
V	GUC	89		43	Y	UAC	60	70	86	R	CGU	18		18
V	GUA	11		5	*	UAA	2		22	R	CGC	45		46
V	GUG	89	209	42	*	UAG	4		44	R	CGA	10		10
S	UCU	7		4	*	UGA	3	9	33	R	CGG	21		21
S	UCC	37		22	H	CAU	6		24	R	AGA	0		0
S	UCA	7		4	H	CAC	19	25	76	R	ACG	5	99	5
S	UCG	59		35	Q	CAA	18		18	G	GGU	74		25
S	AGU	7		4	Q	CAG	83	101	82	G	GGC	160		54
S	AGC	52	169	31						G	GGA	22		7
										G	GGG	43	299	14

TOTAL 2618

The table is based on the codon usage of nine genes from the *M. tuberculosis* complex, starting from the most likely translation initiation codon as identified in Figure 8.3. Similar or identical sequences (64/65K, 10–12K, 19–22K) have been included as one gene. The number of ocurrences of each codon is shown, together with the totals of each amino acid and the percentages of these totals represented by each codon.

reflects the different levels of the relevant tRNA species that are specific for the same amino acid but with different codon specificity.

With mycobacteria, all genes sequenced so far show a relatively high degree of codon bias, which is reflected in a preponderance of those codons with G or C at position 3. This effect has been commented on by many authors in respect of individual genes, and it is apparent from the cumulated codon usage tables for the *M. tuberculosis* complex (Table 8.2) and *M. leprae* (Table 8.3). The effect is much more pronounced for *M. tuberculosis* than for *M. leprae*, which is consistent with the lower G+C content of the DNA of the latter. For example, of the six available serine codons, the three 'G/C' codons UCC, UCG, AGC account for 88% of the serine residues in *M. tuberculosis* genes, while in *M. leprae* the distribution of serine codons is comparatively even apart from the low level of AGU codons.

However, these tables also reveal some more specific features, such as the almost complete absence of the AGA/AGG codons for arginine, the former of which has been recorded just once in *M. leprae* and not at all in *M. tuberculosis*. Of the other arginine codons, the G/C-rich triplet CGG does not occur significantly more often than CGU in *M. tuberculosis*, and actually less often in *M. leprae*. In both species, the favoured arginine codon is CGC by a considerable margin, accounting for nearly half of all arginine residues.

Similarly, with glycine one triplet GGC accounts for about half of all the occurrences of glycine codons, with the next most common being GGU rather than GGG. It is interesting to note that GGA and GGG are classed as 'rare' codons in *E. coli* (Ikemura, 1981, 1985; Sharp and Li, 1986).

It is to be expected that this pattern of codon usage will be a reflection of the distribution of tRNA species in mycobacteria. The existence of such a high degree of codon bias may also pose some interesting evolutionary problems with mycobacteria that may result in an unusually high degree of conservation of the DNA sequence of specific genes, i.e. there will be constraints on the degree of 'silent' mutations that will be consistent with maintaining an acceptable level of expression of the gene concerned (Ikemura, 1985). This may contribute to the observed sequence identity between *M. tuberculosis* and *M. bovis* or BCG in, for example, the 64/65K, 19–22K and 10–12K antigen genes (see Table 8.1 for references).

It is usual to be able to identify possible coding regions by the existence of open reading frames (ORFs) in which there are no stop codons present. In mycobacteria, this could be rendered more difficult since the stop codons are A/U-rich and will therefore occur less frequently. The scale of this problem should not be exaggerated: a random DNA sequence of 70% G+C will have an average ORF size of about 150 bases; this is confirmed by analysis of those gene sequences available which show that in the non-coding reading frames the largest ORF regions are commonly between 100 and 300 bases, and can

Table 8.3 Codon Usage in *Mycobacterium leprae*.

a.a.	codon	number	tot	%	a.a.	codon	number	tot	%	a.a.	codon	number	tot	%
F	UUU	7		20	P	CCU	17		17	N	AAU	11		22
F	UUC	28	35	80	P	CCC	18		18	N	AAC	38	49	78
					P	CCA	17		17					
L	UUA	1		1	P	CCG	47	99	47	K	AAA	19		25
L	UUG	25		20						K	AAG	57	76	75
L	CUU	11		9	T	ACU	15		15					
L	CUC	22		17	T	ACC	45		46	D	GAU	23		29
L	CUA	13		10	T	ACA	15		15	D	GAC	57	80	71
L	CUG	56	128	44	T	ACG	23	98	23					
										E	GAA	27		31
I	AUU	17		23	A	GCU	28		18	E	GAG	60	87	69
I	AUC	49		66	A	GCC	66		43					
I	AUA	8	74	11	A	GCA	19		13	C	UGU	4		44
					A	GCG	39	152	26	C	UGC	5	9	56
M	AUG	26	26	100										
					Y	UAU	9		26	W	UGG	14	14	100
V	GUU	12		11	Y	UAC	25	34	74					
V	GUC	52		47						R	CGU	13		27
V	GUA	12		11	*	UAA	1		20	R	CGC	23		47
V	GUG	34	110	31	*	UAG	1		20	R	CGA	3		6
					*	UGA	3	5	60	R	CGG	8		16
S	UCU	14		19						R	AGA	1		2
S	UCC	14		19	H	CAU	3		19	R	AGG	1	49	2
S	UCA	12		16	H	CAC	13	16	81					
S	UCG	15		20						G	GGU	45		35
S	AGU	3		4	Q	CAA	12		27	G	GGC	61		48
S	AGC	16	74	22	Q	CAG	32	44	73	G	GGA	8		6
										G	GGG	13	127	10
													TOTAL	1386

The table is based on the codon usage of five genes from *M. leprae*, starting from the most likely translation initiation codon as identified in Figure 8.3. The number of occurrences of each codon is shown, together with the totals of each amino acid and the percentages of these totals represented by each codon.

therefore easily be distinguished from the correct reading frame for all but the smallest genes. It is nevertheless useful to support this analysis by one of the many programs that are available for testing the likelihood of an ORF coding for a protein product, on positional base preference or codon usage tables such as those presented in Tables 8.2 and 8.3. In order to avoid circularity of argument, ORFs that have been identified in this way, without evidence of an actual product, have not been included in the codon usage tables in this chapter.

3.2 Post-translational modification

There are a wide variety of possible events that can occur after a protein has been translated, and indeed may well need to occur if a biologically active product is to be formed. Some of these post-translational modifications are concerned with protein secretion, while others affect the final conformation of the protein, are necessary for enzyme function, or are involved with controlling enzyme activity. These events are significant both in understanding the nature of the expression and function of specific products in mycobacteria, and also because of their effect on expression of mycobacterial genes in foreign host where these post-translational events may not occur in the correct fashion. However, few of these events have yet been characterized in mycobacteria.

A number of mycobacterial antigens are present in culture filtrates, and some of these are thought to be genuinely secreted proteins. Protein secretion in bacteria generally requires the presence of a signal peptide at the N-terminus, which is cleaved after secretion to release the mature form of the protein into the supernatant. The essential features of a signal sequence have been summarized as (i) one or more positively charged amino acids at the extreme amino terminus, (ii) an unbroken stretch of eight or more hydrophobic or neutral amino acids with a strong tendency to form an alpha-helix, and (iii) the presence of a signal peptidase cleavage site (Pugsley and Schwartz, 1985).

The existence of a signal peptide is usually inferred from a discrepancy between the experimentally determined N-terminal amino acid sequence of the mature protein and the likely translational start position as determined by DNA sequence analysis. Since the real translational start codon is very rarely known, and there are often several possible alternatives, the length and nature of the signal peptide is often rather uncertain, and indeed some of those claimed for various mycobacterial genes do not conform very well to the above guidelines. Young *et al.* (this volume) provide a list of secreted mycobacterial proteins and possible signal peptide sequences.

A post-translational modification of a different nature that has been studied to some extent in mycobacteria is the addition of biotin. Biotin is an essential cofactor for enzymes involved in carboxylation reactions, primarily the acyl CoA carboxylase(s) involved in fatty acid synthesis. The unusual feature of biotin in comparison with other coenzymes is that it is covalently attached to the enzymes concerned, via the e-NH$_2$ group of a lysine residue.

In *E. coli*, a single biotin carrier protein (BCP) accounts for over 90% of the protein-bound biotin; this is a dimeric protein with a monomer molecular weight of 22K (Fall and Vagelos, 1972), and is often clearly visible on Western blots developed with a streptavidin detection system (Collins *et al.*, 1987). In *E. coli*, the complete acyl CoA carboxylase is a larger complex consisting of a biotin carboxylase and carboxytransferase in addition to the BCP. A propionyl CoA carboxylase from *M. smegmatis* was isolated by Henrikson and Allen (1979) and shown to contain two subunits of 64K and 57K, with biotin being associated exclusively with the heavier subunit. An earlier report (Erfle, 1973) probably refers to the same enzyme. Mycobacteria appear to be a particularly prolific source of biotinylated protein, which may be significant in view of the proposed involvement of biotinylated enzymes in the synthesis of characteristically mycobacterial products such as the mycolic acids (Takayama and Qureshi, 1984). The genes responsible for the biotinylated proteins of *M. leprae*, *M. bovis* and BCG have been cloned and expressed in *E. coli* (Collins *et al.*, 1987) and further characterization of these genes and their products is in progress. The gt11 clone Y3184, from an *M. leprae* gene library, (Young *et al.*, 1985) also expresses a biotinylated protein, as a beta-galactosidase fusion product (Moss, 1987).

It is interesting that we were not only able to get expression of the genes for the mycobacterial BCPs from their own transcriptional and translational control signals, but also the products were biotinylated, resulting in a band on the Western blots in an equivalent position to that seen with mycobacterial extracts. However it is known that the site of biotinylation is highly conserved, consisting of the sequence Ala-Met-Lys-Met (Wood and Barden, 1977). It is therefore to be anticipated that the same sequence occurs in the mycobacterial BCPs, thus accounting for the apparently accurate biotinylation of these proteins in the recombinant *E. coli* cells.

4 CONCLUSION

Although in most cases direct confirmation of the regulatory processes is lacking in mycobacteria, sufficient information is now available, through gene cloning and sequencing, to infer that most of the systems will be found to be essentially similar to those operating in other bacteria. This should come as no surprise, but it does indicate that the current attempts to express foreign antigens in mycobacterial species may be expected to encounter no obstacles more serious than those existing in other organisms. The most interesting aspect of gene regulation in mycobacteria may well prove to be the disclosure of the impact that intracellular growth has on gene expression. We may well anticipate that those genes that are selectively expressed inside macrophages (for example) could play a disproportionate role in the pathogenicity and immunogenicity of the organism.

REFERENCES

Amemura, M., Makino, K., Shinagawa, H., Kobayashi, A. and Nakata, A. (1985). *J. Mol. Biol.* **184**, 241–50.

Andersen, A. B. and Hansen, E. B. (1989). *Infect. Immun.* **57**, 2481–8.

Andersen, A. B., Worsaae, A. and Chaparas, S. D. (1988). *Infect. Immun.* **56**, 1344–51.

Ashbridge, K. R., Booth, R. J., Watson, J. D. and Lathigra, R. B. (1989). *Nucleic Acids Res.* **17**, 1249.

Bagg, A. and Neilands, J. B. (1987). *Microbiol. Rev.* **51**, 509–18.

Baird, P. N., Hall, L. M. C. and Coates, A. R. M. (1989). *J. Gen. Microbiol.* **135**, 931–9.

Barclay, R. and Ratledge, C. (1988). *J. Gen. Microbiol.* **134**, 771–6.

Barclay, R. and Wheeler, P. R. (1989). In *The Biology of the Mycobacteria* (eds C. Ratledge, J. Stanford and J. M. Grange), pp. 37–106. Academic Press, London.

Booth, R. J., Harris, D. P., Love, J. M. and Watson, J. D. (1988). *J. Immunol.* **140**, 597–601.

Borremans, M., De Wit, L., Volckaert, G., Ooms, J., De Bruyn, J., Huygen, K., Van Vooren, J.-P., Stelandre, M., Verhofstadt, R. and Content, J. (1989). *Infect. Immun.* **57**, 3123–30.

Bowles, J. A. and Segal, W. (1965). *J. Bacteriol.* **90**, 157–63.

Calderwood, S. B. and Mekalanos, J. J. (1987). *J. Bacteriol.* **169**, 4759–64.

Calderwood, S. B. and Mekalanos, J. J. (1988). *J. Bacteriol.* **170**, 1015–17.

Cherayil, B. J. and Young, R. A. (1988). *J. Immunol.* **141**, 4370–5.

Choubey, D. and Gopinathan, K. P. (1986). *Curr. Microbiol.* **13**, 171–5.

Clark-Curtiss, J. E., Jacobs, W. R., Docherty, M. A., Ritchie, L. R. and Curtiss, R. (1985). *J. Bacteriol.* **161**, 1093–102.

Collins, M. E., Moss, M. T., Wall, S. and Dale, J. W. (1987). *FEMS Microbiol. Lett.* **43**, 53–6.

Collins, M. E., Patki, A., Wall, S., Nolan, A., Goodger, J., Woodward, M. J. and Dale, J. W. (1990). *J. Gen. Microbiol.* (in press).

Cowing, D. W., Bardwell, J. C. A., Craig, E. A., Woolford, C., Hendrix, R. W. and Gross, C. A. (1985). *Proc. Nat. Acad. Sci. USA* **82**, 2679–83.

de Grandis, S., Ginsberg, J., Toone, M., Climie, S., Friesen, J. and Brunton, J. (1987). *J. Bacteriol.* **169**, 4313–19.

de Lorenzo, V., Wee, S., Hererro, M. and Neilands, J. B. (1987). *J. Bacteriol.* **169**, 2624–30.

Draper, P. (1967). *J. Gen. Microbiol.* **46**, 111–23.

Erfle, J. D. (1973). *Biochimica et Biophysica Acta* **316**, 143–55.

Fall, R. R. and Vagelos, P. R. (1972). *J. Biol. Chem.* **247**, 8005–15.

Finkelstein, R. A., Sciortino, C. V. and McIntosh, M. A. (1983). *Rev. Infect. Dis.* **5**, S759–77.

Goodfellow, M. and Minnikin, D. E. (1984). In *The Mycobacteria: a Sourcebook* (eds G. P. Kubica and L. G. Wayne), pp. 1–24. Marcel Dekker, New York and Basel.

Greenfield, L., Bjorn, M. J., Horn, G., Fong, D., Buck, G. A., Collier, R. J. and Kaplan, D. A. (1983) *Proc. Nat. Acad. Sci. USA* **80**, 6853–7.

Hall, R. M., Sritharan, M., Messenger, A. J. M. and Ratledge, C. (1987). *J. Gen. Microbiol.* **133**, 2107–14.

Hantke, K. (1981). *Mol. Gen. Genet.* **182**, 288–92.

Hantke, K. (1984). *Mol. Gen. Genet.* **197**, 337–41.

Henrikson, K. P. and Allen, S. H. G. (1979). *J. Biol. Chem.* **254**, 5888–91.

Ikemura, T. (1981). *J. Mol. Biol.* **146**, 1–21.

Ikemura, T. (1985). *Mol. Biol. Evol.* **2**, 13–34.

Iwainsky, H. and Sehrt, I. (1970). *Zentralbl. Bakteriol.* **213**, 222–32.

Iwainsky, H. and Sehrt, I. (1971). *Zentralbl. Bakteriol.* **218**, 212–23.

Izumori, K., Ueda, Y. and Yamanaka, K. (1978). *J. Bacteriol.* **133**, 413–14.

Jacobs, W. R., Barrett, J. F., Clark-Curtiss, J. E. and Curtiss, R. (1986a). *Infect. Immun.* **52**, 101–9.

Jacobs, W. R., Docherty, M. A., Curtiss, R. and Clark-Curtiss, J. E. (1986b). *Proc. Nat. Acad. Sci. USA* **83**, 1926–30.

Kaczorek, M., Delpeyroux, F., Chenciner, N. and Streeck, R. E. (1983). *Science* **221**, 855–8.

Kanei, C., Uchida, T. and Yoneda, M. (1977). *Infect. Immun.* **18**, 203–9.

Kasik, J. E. (1979). In *Beta-lactamases* (eds J. M. T. Hamilton-Miller and J. T. Smith), pp. 339–50. Academic Press, London.

Kieser, T., Moss, M. T., Dale, J. W. and Hopwood, D. A. (1986). *J. Bacteriol.* **168**, 72–80.

Lathigra, R. B., Young, D. B., Sweetser, D. and Young, R. A. (1988). *Nucleic Acids Res.* **16**, 1636.

Leong, D. and Murphy, J. R. (1985). *J. Bacteriol.* **163**, 1114–19.

Makino, K., Shinagawa, H., Amemura, M. and Nakata, A. (1986). *J. Mol. Biol.* **190**, 37–44.

Matsuo, K., Yamaguchi, R., Yamazaki, A., Tasaka, H. and Yamada, T. (1988). *J. Bacteriol.* **170**, 3847–54.

Mehra, V., Sweetser, D. and Young, R. A. (1986). *Proc. Nat. Acad. Sci. USA* **83**, 7013–17

Messenger, A. J. M. and Ratledge, C. (1982). *J. Bacteriol.* **149**, 131–5.

Moss, M. T. (1987). Cloning and expression of mycobacterial genes in *Escherichia coli*. PhD thesis, University of Surrey.

Murphy, J. R., Skiver, J. and Bride, G. M. (1976). *J. Virol.* **18**, 235–44.

Nash, D. R., Wallace, R. J., Steingrube, V. A., Udou, T., Steele, L. C. and Forrester, G. D. (1986). *Am. Rev. Respir. Dis.* **134**, 1276–82.

Neilands, J. B. (1981). *Ann. Rev. Nutr.* **1**, 27–46.

Pugsley, A. P. and Schwartz, M. (1985). *FEMS Microbiol. Rev.* **32**, 3–38.

Ratledge, C. (1982). In *The Biology of the Mycobacteria*, Vol. 1 (eds C. Ratledge and J. Stanford), pp. 186–241. Academic Press, New York and London.

Ratledge, C. (1984). In *The Mycobacteria: a Sourcebook* (eds G. P. Kubica and L. G. Wayne), pp. 603–27. Marcel Dekker, New York and Basel.

Ratti, G., Rappuoli, R. and Giannini, G. (1983). *Nucleic Acids Res.* **11**, 6589–95.

Segal, W. and Edwards, B. S. (1984). In *The Mycobacteria: a Sourcebook* (eds G. P. Kubica and L. G. Wayne), pp. 575–94. Marcel Dekker, New York and Basel.

Sharp, P. M. and Li, W-H. (1986). *Nucleic Acids Res.* **14**, 7737–49.

Shinnick, T. M. (1987). *J. Bacteriol.* **169**, 1080–8.

Shinnick, T. M., Plikaytis, B. P., Hyche, A. D., van Landingham, R. M. and Walker L. V. (1989). *Nucleic Acids Res.* **17**, 1254.

Snapper, S. B., Lugosi, L., Jekkel, A., Melton, R. E., Kieser, T., Bloom, B. R. and Jacobs, W. R. (1988). *Proc. Nat. Acad. Sci. USA* **85**, 6987–91.

Sokol, P. A., Cox, C. D. and Iglewski, B. H. (1982). *J. Bacteriol.* **151**, 783–7.

Sritharan, M., Wheeler, P. R. and Ratledge, C. (1989). *European J. Biochem.* **180**, 587–93.
Sritharan, V., Wheeler, P. R. and Ratledge, C. (1990). *J. Gen. Microbiol.* **136**, 203–9.
Staden, R. (1984). *Nucleic Acids Res.* **12**, 505–19.
Surin, B. P., Jans, D. A., Fimmel, A. L., Shaw, D. C., Cox, G. B. and Rosenberg, H. (1984). *J. Bacteriol.* **157**, 772–8.
Surin, B. P., Rosenberg, H. and Cox, G. B. (1985). *J. Bacteriol.* **161**, 189–98.
Tai, S-P. S. and Holmes, R. K. (1988). *Infect. Immun.* **56**, 2430–6.
Takayama, K. and Qureshi, N. (1984). In *The Mycobacteria: a Sourcebook* (eds G. P. Kubica and L. G. Wayne), pp. 315–44. Marcel Dekker, New York and Basel.
Terasaka, K., Yamaguchi, R., Matsuo, K., Yamazaki, A., Nagai, S. and Yamada, T. (1989). *FEMS Microbiol. Lett.* **58**, 273–6.
Thangaraj, H. S., Lamb, F. I., Davis, E. O. and Colston, M. J. (1989). *Nucleic Acids Res.* **17**, 8378.
Thole, J. E. R., Dauwerse, H. G., Das, P. K., Groothuis, D. G., Schouls, L. M. and van Embden, J. D. A. (1985). *Infect. Immun.* **50**, 800–6.
Thole, J. E. R., Keulen, W. J., Kolk, A. H. J., Groothuis, D. G., Berwald, L. G., Tiesjema, R. H. and van Embden, J. D. A. (1987). *Infect. Immun.* **55**, 1466–75.
Thole, J. E. R., Stabel, L. F. E. M., Suykerbuyk. M. E. G., de Wit, M. Y. L., Klatser, P. R., Kolk, A. H. J. and Hartskeerl, R. A. (1990). *Infect. Immun.* **58**, 80–7.
Tommassen, J., Koster, M. and Overduin, P. (1987). *J. Mol. Biol.* **198**, 633–41.
Wood, H. G. and Barden, R. E. (1977). *Ann. Rev. Biochem.* **46**, 385–413.
Yamaguchi, R., Matsuo, K., Yamazaki, A., Nagai, S., Terasaka, K. and Yamada, T. (1988). *FEBS Lett.* **240**, 115–17.
Yamaguchi, R., Matsuo, K., Yamazaki, A., Abe, C., Nagai, S., Terasaka, K. and Yamada, T. (1989). *Infect. Immun.* **57**, 283–8.
Young, R. A., Bloom, B. R., Grosskinsky, C. M., Ivanyi, J., Thomas, D. and Davis, R. W. (1985). *Proc. Nat. Acad. Sci. USA* **82**, 2583–7.
Zainuddin, Z., Kunze, Z. and Dale J. W. (1989). *Mol. Microbiol.* **3**, 29–34.

9 Molecular genetic approaches to mycobacterial investigation

Scott B. Snapper, Barry R. Bloom and
William R. Jacobs, Jr

*Department of Microbiology and Immunology,
Albert Einstein College of Medicine, Bronx, NY 10461, USA*

1 INTRODUCTION

The application of the recombinant DNA and monoclonal antibody technology to mycobacteria has brought forth an explosion of studies investigating mycobacterial genes and their protein products. Over the past five years, the genomes of several mycobacterial pathogens, including *M. leprae* and *M. tuberculosis*, have been cloned in *E. coli* using both plasmid and phage expression vectors (Clark-Curtiss *et al.*, 1985; Young *et al.*, 1985a, b; Thole *et al.*, 1985; Jacobs *et al.*, 1986; Khandekar *et al.*, 1986). The expression of mycobacterial genes in *E. coli* has permitted the isolation and detailed characterization of mycobacterial proteins which stimulate the human and murine immune response upon mycobacterial infection (see chapters by D. B. Young, J. E. R. Thole, and M. J. Colston, this volume). While these analyses depended on the ability to express mycobacterial genes in *E. coli*, it seemed likely that many mycobacterial genes would not be expressed since the *E. coli* transcriptional and translational apparatus might not recognize some mycobacterial expression sequences (see Dale and Patki, this volume). Clark-Curtiss *et al.* (1985) first demonstrated the difficulty in expressing mycobacterial genes in *E. coli* from mycobacterial promoters; *M. leprae* cosmid libraries failed to complement nine different auxotrophic markers. To overcome these limitations, mycobacterial genes were cloned in vectors under the control of *E. coli* expression signals (Young *et al.*, 1985a, b; Thole *et al.*,

MOLECULAR BIOLOGY OF THE MYCOBACTERIA
ISBN 0–12–483378–0

1985; Jacobs *et al.*, 1986; Khandekar *et al.*, 1986). One extraordinarily successful approach was developed by constructing *E. coli*–mycobacterial gene fusions allowing for the production of hybrid proteins in a bacteriophage lambda expression vector (Young *et al.*, 1985a, b). Mycobacterial proteins were then identified using mycobacterial specific antisera. Another strategy taken was to clone *M. leprae* genes into a plasmid expression vector under the strong *asd* promoter from *Streptococcus mutans*. Several clones were detected which succeeded in complementing a citrase synthase auxotroph, establishing that at least some mycobacterial enzymes can function in *E. coli* (Jacobs *et al.*, 1986). An *M. bovis* antigen was also expressed in *E. coli* and identified with antiserum when BCG DNA was cloned downstream of an *E. coli* promoter in a lambda vector (Thole *et al.*, 1985).

Another inherent limitation to using *E. coli* as a cloning host is that certain antigens which require complex biosynthetic pathways (e.g. polysaccharide, lipids) might not be formed. Since these antigens may be integral to the method(s) by which mycobacterial pathogens invade and establish a niche in their hosts, some investigators have sought other hosts for the expression of mycobacterial genes (Kieser *et al.*, 1986). *Streptomyces lividans* is an ideal alternative host because cloning strategies are well worked out (Hopwooe *et al.*, 1985) and because of its relatedness to the mycobacteria, both containing high glycine plus cytosine DNA (72%, 63%, and 56% for *S. lividans*, *M. tuberculosis* and *M. leprae*, respectively). Since earlier studies had shown that mycobacterial transcriptional signals did not function well in *E. coli* (Clark-Curtiss *et al.*, 1985; Thole *et al.*, 1985), promoter probe constructs were used to determine that mycobacterial DNA could efficiently direct the *S. lividans* transcriptional machinery (Kieser *et al.*, 1986).

While alternative cloning hosts continue to be effective in studying mycobacterial genes, an efficient gene transfer system in mycobacteria would greatly increase the potential of molecular genetic approaches for the study of mycobacteria. The development of such genetic systems, allowing for the transfer, mutation, and expression of foreign genes in mycobacteria, would minimize host-specific constraints in expressing mycobacterial genes. Nonpathogenic strains of mycobacteria, such as *M. smegmatis* and BCG, could be used as cloning hosts for genes from the virulent mycobacteria. This approach would allow for the development of novel strategies to identify and characterize virulence genes and the expression of antigens requiring complex biosynthetic pathways. Avirulent mycobacterial strains, moreover, could serve as cloning hosts for genes encoding potential chemotherapeutic targets. With efficient mycobacterial gene transfer systems, it is feasible to manipulate directly the pathogenic mycobacteria to identify virulence genes or protective antigens which could be used to develop more effective mycobacterial vaccines. Since BCG is currently one of the most widely used

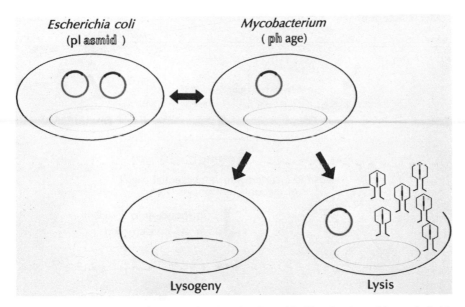

Figure 9.2 Schematic diagram of a shuttle phasmid. Shuttle phasmids are hybrid molecules which replicate as plasmids in *E. coli* and as phage in mycobacteria. Virulent shuttle phasmids lyse their hosts, whereas temperate shuttle phasmids can integrate within the mycobacterial chromosome and replicate stably as prophage.

replication, into which foreign DNA can be inserted. Furthermore, any strategy to develop a phage-based genetic system must include mechanisms to identify any nonessential regions of the phage, to delete such regions to optimize cloning capacity, and to allow for efficient cloning of foreign DNA into unique restriction sites.

3.1   Construction of a shuttle phasmid

Jacobs *et al.* (1987) developed a phage vector which permitted both the manipulation and amplification of mycobacterial DNA constructs in *E. coli*, and subsequent transfer and replication in a variety of mycobacteria. These novel recombinant DNA vectors were termed 'shuttle phasmids' because they replicate in mycobacteria as phage and in *E. coli* as plasmids (Figure 9.2). The construction of these recombinant mycobacteriophage was facilitated, fortuitously, by the observation that mycobacterial DNA was not expressed well in *E. coli* (Clark-Curtiss *et al.*, 1985) thus providing conditional expression of mycobacteriophage lytic functions. A requirement of the cloning strategy, which will become readily apparent below, was that the

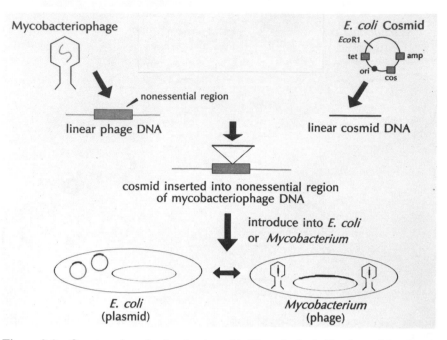

Figure 9.3   Construction of a shuttle phasmid. The principal objectives of the shuttle phasmid construction are illustrated (for details of construction see text and Jacobs *et al.*, 1987, 1989). An *E. coli* cosmid is inserted into a nonessential region of a mycobacteriophage DNA molecule and introduced into *E. coli* or mycobacteria. Shuttle phasmids replicate as plasmids in *E. coli* and as phage in mycobacteria.

phage used be approximately 50 kb. The initial phage chosen for vector development was TM4, a 53 kb temperate phage isolated from *M. avium* (Timme and Brennan, 1984).

In the shuttle phasmid cloning strategy (schematized in Figure 9.3 and described in detail elsewhere by Jacobs *et al.*, 1987, 1989), complete *E. coli* cosmids (Collins and Hohn, 1978) — i.e. plasmids containing the cohesive end (*cos*) of the bacteriophage lambda, an *E. coli* origin of replication (*ori*), unique restriction sites, and selectable markers — are inserted randomly throughout the mycobacteriophage TM4 genome. This was accomplished by digesting ligated phage DNA with a frequent cutting restriction enzyme (e.g. *Sau*3A) at very dilute concentrations such that random cuts were obtained at approximately every 30–50 kb. Cosmids, cut singly with a compatible restriction enzyme (e.g. *Bam*H1), are then ligated to this phage DNA. The final product consisted of large, essentially random, fragments of phage DNA flanked by cosmid DNA. The lambda *cos* site within the cosmid allows for selective cloning of large DNA fragments using a lambda *in vitro* packaging system. To package recombinant DNA into phage lambda heads, this system requires two *cos* sites to be separated by approximately 50 kb. Since a cosmid can be as small as 4 kb, DNA fragments of 42–50 kb can be efficiently cloned into cosmid vectors, which is why it is necessary to use a phage of approximately 50 kb.

Following *in vitro* packaging and subsequent lambda phage transduction, recombinant cosmid–mycobacteriophage molecules replicate as plasmids in *E. coli*. Transductants are selected for by their resistance to an antibiotic resulting from expression of a selectable marker contained on the cosmid. The 50 kb supercoiled plasmid DNA molecules isolated from transductants represent a library of recombinant molecules of TM4 phage DNA into which a cosmid had been randomly inserted in sites around the TM4 genome.

Recombinant phage molecules with cosmids inserted in nonessential regions of the TM4 genome were identified by transfecting *M. smegmatis* protoplasts with the supercoiled library and then searching for plaque forming units. Those recombinant molecules containing cosmid molecules in a nonessential region of the mycobacteriophage genome would replicate in *M. smegmatis* as phage and by definition would be shuttle phasmids, whereas those molecules into which the cosmid had been inserted in essential genes would not yield viable phage.

A TM4-shuttle phasmid, phAE1, was isolated from a plaque and investigated further. DNA isolated from phAE1 could be re-introduced into *E. coli*, where it replicated as a plasmid. Alternatively, intact phAE1 phage particles could be re-introduced into *M. smegmatis* or introduced into other strains of mycobacteria, including BCG, efficiently by simple infection. An efficient method of introducing recombinant molecules into the slow-growing mycobacterial strains, which is likely to be extended to include the pathogenic mycobacteria, is arguably the single greatest asset of this shuttle phasmid strategy (see Section 3.5).

3.2   Construction of a temperate shuttle phasmid

Having established that shuttle phasmids could be used to propagate efficiently foreign DNA in the mycobacteria, similar vectors were required which stably maintained foreign DNA in the mycobacterial cell. Although TM4 had been reported to lysogenize *M. avium*, stable *M. smegmatis* or BCG lysogens were not obtained with phAE1. One attractive explanation was pseudolysogeny, as several accounts of unstable lysogens, or pseudolysogens, have been reported to arise following infection with some putative temperate mycobacteriophages (Baess, 1970, 1971). Snapper *et al.* (1988) extended the shuttle phasmid methodology by developing phasmids from a different temperate mycobacteriophage. First they confirmed early reports (Doke,

1960; Tokunaga and Sellers, 1970b) that the mycobacteriophage L1 produced turbid plaques on *M. smegmatis* and conferred resistance to L1 superinfection upon lysogens. Then, these investigators demonstrated by Southern analysis that, concomitant with lysogenization, L1 integrates site-specifically into the *M. smegmatis* chromosome. Shuttle phasmids were subsequently constructed with L1 in a manner similar to those phasmids constructed from the TM4 phage. Phasmids were isolated which like TM4-shuttle phasmids replicate in *E. coli* as plasmids, and in *M. smegmatis* as phage; however, L1–phasmids, like the parent phage, form stable lysogens integrating into the mycobacterial chromosome upon lysogenization. These temperate shuttle phasmids thus provided for stable maintenance of recombinant DNA molecules in *M. smegmatis*. And like the TM4–shuttle phasmids, L1–shuttle phasmids can infect BCG.

3.3 Cloning genes into shuttle phasmids

TM4– and L1–shuttle phasmids have unique restriction sites within the cosmid portion of the molecule allowing for the cloning of foreign genes (Jacobs *et al.*, 1987, 1989; Snapper *et al.*, 1988). The lambda cohesive ends enables efficient cloning of genes with large DNA molecules using standard cosmid cloning strategies (Collins and Hohn, 1978). (See review by Jacobs *et al.*, 1989, for a more detailed explanation of the use of cosmid cloning with shuttle phasmids.) The plasmid form of the shuttle phasmid can be digested with a unique restriction enzyme and ligated to a gene of interest at a concentration resulting in concatemers. This DNA can be subsequently packaged into lambda heads using standard *E. coli in vitro* packaging extracts and introduced efficiently into *E. coli* and later into mycobacterial strains.

3.4 Expression of a selectable marker gene in mycobacteria

The ability to introduce recombinant DNA stably into mycobacteria cells using a temperate L1–shuttle phasmid provided a means for testing the expression of foreign genes in mycobacteria. Snapper *et al.* (1988) inserted a gene cartridge containing the Tn*903* gene encoding aminoglycoside phosphotransferase, which confers kanamycin-resistance in *E. coli*, into a unique restriction site within the L1-shuttle phasmid, phAE15, by the efficient cosmid cloning techniques just described. A recombinant shuttle phasmid containing the selectable marker, phAE19, was first isolated by its ability to confer resistance to kanamycin to *E. coli* transductants. phAE19 was subsequently introduced into *M. smegmatis* protoplasts by transfection, and found to integrate into the *M. smegmatis* chromosome. Most importantly, these shuttle phasmids conferred kanamycin-resistance upon lysogens, establishing kanamycin-resistance as the first selectable marker expressed in mycobacteria. Preliminary studies have furthermore suggested that L1-shuttle phasmids may confer kanamycin-resistance upon BCG cells (S. B. Snapper, and W. R. Jacobs, unpublished results).

3.5   Shuttle phasmids as multifunctional vectors

Shuttle phasmid DNA can be packaged into bacteriophage lambda particles by *in vitro* cosmid cloning techniques where they can be efficiently introduced into *E. coli* to replicate as plasmids. Alternatively, shuttle phasmid DNA introduced into mycobacterial protoplasts by transfection can replicate as phage, lysing or lysogenizing its host depending on the specific phage type. Recombinant mycobacteriophage particles can subsequently be isolated and introduced into a variety of strains of mycobacteria by infection, obviating the necessity of protoplast production. Circumventing a protoplast preparation step may be essential to introduce efficiently recombinant DNA into the slow-growing mycobacterial strains. Foreign genes can now be cloned and manipulated in shuttle phasmids in *E. coli* and then introduced into mycobacteria. These vectors will provide one approach to the study of mycobacterial pathogenesis, allowing for the cloning of genes from the virulent mycobacteria in nonpathogens and thereby providing methods to isolate or test putative virulence genes (see Section 5). Temperate shuttle phasmids carrying protective antigens from a variety of pathogens may also be used to develop BCG into a recombinant mycobacterial multivaccine vehicle (see Section 6).

3.6   Further applications for phages in the development of genetic systems

Mycobacteriophages have the potential to be exploited for a variety of other practical tools for use in mycobacterial genetic analys s. *In vitro* packaging systems, similar to the systems described above for *E. coli* (Rosenberg *et al.*, 1985), can be developed from mycobacteriophages providing highly efficient methods of introducing recombinant DNA directly into mycobacteria by phage infection. Temperate phages, like L1, could also provide site-specific integration sequences which can be used to introduce permanently genes into a mycobacterial chromosome. This may be particularly useful in the development of a recombinant BCG multivaccine (Section 6). Additionally, phages could be used as a source of strong regulatable promoters for expression of foreign or potentially lethal genes in mycobacteria, and as a source of replicons for the development of a plasmid-based genetic system.

4 PLASMID-BASED GENE TRANSFER SYSTEMS

Plasmid-based genetic systems can extend the capabilities of phage-based systems by offering particular advantages such as increased cloning capacities, ease of DNA manipulation, and copy number control. The development of a plasmid-based genetic system in mycobacteria would require an efficient mechanism of introducing plasmid DNA into mycobacterial cells and a plasmid which contains a functional mycobacterial origin of replication, and a selectable marker.

4.1 Mycobacterial plasmids

Plasmids were first described in mycobacteria by Crawford and Bates (1979) in *M. avium-intracellulare*. Other investigators have since identified plasmids in *M. fortuitum* (Hull *et al.*, 1984; Labidi *et al.*, 1984), and the *M. avium-intracellulare–scrofulaceum* complex (MAIS) (Crawford and Bates, 1979; Crawford *et al.*, 1981b; Mizuguchi *et al.*, 1981; Scott Meissner and Falkinham, 1986), whereas plasmids have yet to be isolated from *M. smegmatis*, *M. tuberculosis*, or BCG. Characterizations of some of the mycobacterial strains containing plasmids have suggested that these plasmids may encode important functions related to colonial morphology (Mizuguchi *et al.*, 1981), antibiotic sensitivity (Mizuguchi *et al.*, 1981, 1983), specific growth requirements (e.g. media supplements, optimal temperature) (Fry *et al.*, 1986) resistance to heavy metal toxicity (Scott Meissner and Falkinham, 1984; Fry *et al.*, 1986), and a restriction-modification system (Crawford *et al.*, 1981a). (For further information regarding plasmids isolated from mycobacterial strains, especially clinical isolates see the chapters by Crawford *et al.* and Martin *et al.*, this volume.)

Two mycobacterial plasmids, pLR7 and pAL5000, originally isolated from *M. fortuitum* (Labidi *et al.*, 1984) and *M. intracellulare* (Crawford *et al.*, 1981b) respectively, were analysed further in *E. coli*. In order to assure replication in *E. coli*, hybrid mycobacteria/*E. coli* plasmids were constructed by inserting pBR322, and *E. coli* plasmid, into pLR7 (Crawford and Bates, 1984) and pAL5000 (Labidi *et al.*, 1985). Following molecular analysis of these hybrid plasmids in *E. coli*, restriction maps of pLR7 (Crawford and Bates, 1984) and pAL5000 (Labidi *et al.*, 1985) were established and recently the complete nucleotide sequence of pAL5000 was determined (Rauzier *et al.*, 1988; see Martin *et al.*, this volume).

4.2 Plasmid transformation in mycobacteria

Snapper *et al.* (1988) developed a different strategy for constructing mycobacteria/*E. coli* hybrid plasmids with the intention of introducing these

plasmids into mycobacteria. To ensure a functional replicon for mycobacteria, they constructed a hybrid plasmid library in *E. coli* by randomly
inserting the *E. coli* plasmid, pIJ666 (Kieser and Melton, 1988), into
pAL5000. The *E. coli* plasmid contained a kanamycin-resistance gene from
Tn5 which was expected to function in mycobacteria because a similar gene
had been expressed in *M. smegmatis* when carried on a temperate shuttle
phasmid (Snapper *et al.*, 1988) (see Section 3.4). Transformation of the
pIJ666::pAL5000 library into *M. smegmatis* protoplasts failed, presumably
due to the difficulty in regenerating viable cells from protoplasts. A report
describing a new method for transforming the Gram-positive prokaryote
Lactobacillus casei by high voltage electroporation, which had been frequently
used to introduce DNA into eukaryotic cells, suggested to these investigators
an alternative approach to bacterial transformation. The attractive feature of
this approach was that whole bacterial cells rather than protoplasts could be
used, thereby obviating the need for a protoplast regeneration step. DNA
was therefore introduced directly into intact *M. smegmatis* cells by electroporation. Conditions were initially developed for electroporation using D29
phage DNA to monitor uptake, yielding $>5 \times 10^3$ plaque-forming units per
microgram of D29 DNA. By using optimal conditions, electroporation of the
pIJ666::pAL5000 DNA library into *M. smegmatis* yielded 1–10 kanamycin-
resistant transformants per microgram (Figure 9.4) and plasmid isolated
from these mycobacterial transformants could retransform *E. coli* at high
efficiency. Furthermore, these shuttle plasmids were introduced successfully
by electroporation into slow-growing mycobacterial BCG substrains with
efficiencies greater than 10^3 kanamycin-resistant transformants obtained per
microgram of plasmid DNA (Figure 9.4) (Snapper *et al.*, 1988; Lugosi *et al.*,
1989). Others (Gicquel-Sanzey *et al.*, 1989) confirmed these experiments
in a report describing the introduction of a similar pAL5000 derived
mycobacteria/*E. coli* shuttle plasmid into *M. smegmatis* and BCG by
electroporation.

   Zainuddin *et al.* (1989a) proposed that pIJ666, which contains a
chloramphenicol-resistance gene in addition to the kanamycin-resistance
gene, can be introduced into *M. smegmatis* protoplasts and will replicate as an
episome independent of any mycobacterial plasmid sequences. Curiously,
pIJ666 *M. smegmatis* transformants expressed chloramphenicol resistance but
failed to express resistance to kanamycin, although DNA isolated from such
transformants could be introduced into *E. coli* conferring both kanamycin-
and chloramphenicol-resistance upon transformants. Moreover, plasmids
isolated from *E. coli* transformants were said to be identical to pIJ666. A
recent report (Zainuddin *et al.*, 1989b), however, indicated that some plasmids
isolated from a putative pIJ666 *M. smegmatis* transformant had picked up
some additional DNA sequences. One intriguing hypothesis consistent with

	transformants / μg plasmid DNA
E. coli	$>10^8$
M. smegmatis (mc^2 6)	0 - 10
M. smegmatis (mc^2 155)	$>10^5$
M. bovis - BCG	$>10^3$

Figure 9.4 Mycobacterial plasmid electro-transformation. Electroporation facilitates the uptake of plasmid DNA by mycobacterial cells. Transformation efficiencies for several mycobacterial strains, and *E. coli* are listed. *M. smegmatis* (mc^{2}6) is a single colony isolate of ATCC607 (Jacobs *et al.*, 1987). *M. smegmatis* (mc^{2}155) is a high efficiency transformation mutant (Snapper *et al.*, 1990).

both reports is that *E. coli* plasmids can integrate into the mycobacterial chromosome, and at some measurable frequency will recombine out and, in the process, remove some chromosomal DNA. Preliminary Southern analysis data suggests that pIJ666 can integrate into the *M. smegmatis* chromosome (S. B. Snapper and W. R. Jacobs, unpublished results).

4.3 Isolation of high efficiency transformation mutants

M. smegmatis may be an ideal host for genetic and molecular analysis of mycobacterial genes as it is fast growing, avirulent, and can be grown on well defined minimal medium. The *M. smegmatis* transformation efficiencies described above — less than 10 transformants per microgram — however preclude most forms of genetic analysis. As a result, high efficiency *M. smegmatis* transformation mutants were isolated which can be transformed by electroporation at efficiencies greater than 10^5 transformants per microgram (Figure 9.4) (Snapper *et al.*, 1990). These mutants have subsequently enabled further analysis of mycobacterial genes, including the mapping of an essential replication region of pAL5000 and the expression of the *M. leprae* gene encoding the 65 kD heat-shock gene in *M. smegmatis* (Snapper *et al.*, 1990).

5   APPLYING MOLECULAR GENETICS TO MYCOBACTERIA

The ability to introduce recombinant DNA into mycobacteria has opened up a myriad of novel prospects for means of understanding the biology of the mycobacteria. The vectors and methodologies described so far provide the basic tools necessary to introduce recombinant molecules into both the slow-growing and fast-growing mycobacteria. It should now be possible to study the basic mechanisms of DNA replication, gene transcription and translation, and gene regulation directly in mycobacterial cells. The application of useful reporter probes and the development of insertional mutagenesis systems should enable characterization of a variety of genes. The identification of genes encoding mycobacterial virulence determinants should help to delineate the key steps of mycobacterial infection leading to disease, as well as to establish those genetic functions necessary for BCG to elicit an immune response. Such basic knowledge will be essential in producing better anti-mycobacterial vaccines and in developing BCG into a recombinant multivalent vaccine vehicle (Jacobs et al., 1987; Snapper et al., 1988). Furthermore, it should now be possible to identify definitively the targets of existing drugs and to develop strategies to identify novel anti-mycobacterial agents and thereby provide novel therapies for treating mycobacterial disease. Clearly, an exciting new chapter in the history of mycobacteria has begun and will continue to develop in the years to come.

5.1   Molecular genetic studies to probe mycobacterial pathogenesis

Studies of molecular pathogenesis of diverse bacteria have grown exponentially in recent years. There are two major experimental themes used to study pathogenesis utilizing recombinant DNA technologies and existing animal models (see review by Isberg and Falkow (1985a)). One strategy involves screening libraries of mutants generated with an insertional mutagenic agent, such as a transposon, for mutants that are now attenuated for their virulent phenotype. Such a strategy has been used to define virulence genes of a wide variety of pathogens, including *Bordatella pertussis* (Weiss et al., 1983), *Shigella flexneri* (Maurelli and Curtiss, 1984), *Yersinia pestis* (Goguen et al., 1984), and *Salmonella typhimurium* (Fields et al., 1986). Although this strategy can be laborious because it involves screening thousands of mutants, the process has been expedited in other bacterial systems by randomly inserting transposons containing special reporter genes such as β-galactosidase (Maurelli and Curtiss, 1984) or alkaline phosphatase (Taylor et al., 1989), which when conditionally expressed have been found to be inserted frequently in virulence genes. An alternative strategy involves constructing genomic libraries of virulent organisms in an appropriate vector, introducing these libraries into avirulent organisms, and then selecting for clones expressing a virulent phenotype in animal- or cell-model systems. This approach has been used successfully to clone the outer membrane proteins of *Shigella flexneri* (Maurelli *et al.*, 1985) or *Yersinia pseudotuberculosis* (Isberg and Falkow, 1985b) in *E. coli* conferring the ability to penetrate epithelial cells.

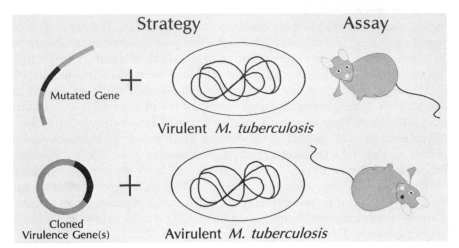

Figure 9.5   Strategies to identify virulence genes in *M. tuberculosis*. In the first strategy (top row) *M. tuberculosis* genes are randomly mutagenized and introduced into virulent *M. tuberculosis* cells to allow for recombination with analogous genes in the chromosome. Mutated virulence genes are identified from *M. tuberculosis* recombinant cells which fail to kill a mouse upon infection. In the second strategy (bottom row), genes from a virulent *M. tuberculosis* strain are cloned on a plasmid vector and introduced into an avirulent strain (e.g. BCG). *M. tuberculosis* virulence genes are identified from recombinant bacteria which can cause a lethal infection in a mouse.

Both of these approaches should be directly applicable to the study of virulence determinants for mycobacteria, such as *M. tuberculosis*, where well characterized mouse or guinea pig models of infection exist (Figure 9.5). Since *M. tuberculosis* will cause a fatal disease in mice, the first strategy would involve the screening of libraries of insertion mutants of *M. tuberculosis* for those that fail to kill a mouse upon host infection. Such an approach requires the development of random mutagenesis systems for mycobacteria that could be generated with an appropriate transposon or an alternative insertional mutagenesis approach. The screening of random libraries of thousands of *M. tuberculosis* mutants in animals, however, would be prohibitively expensive. It may be possible to first screen the *M. tuberculosis* mutants in macrophage model systems. Alternatively, screens that identify a subset of mutants that are likely to include avirulent phenotypes, such as mutants that have aberrant cell walls, might prove useful in identifying virulence genes.

The second strategy for identifying *M. tuberculosis* virulence genes utilizes a selection in animals that may improve its chance of successs. The basic strategy would involve introducing DNA fragments from a virulent *M. tuberculosis* strain, such as H37Rv, into well-characterized attenuated tubercle strains, such as H37Ra or BCG. Pools of libraries of such strains may then be introduced into a mouse and a successful selection would result in a mouse fatality from a normally attenuated strain. Genes affecting virulence would then be identified by sequence analysis and subsequent characterization. Once identified, the study of the gene products of such virulence genes could lead to the development of a better vaccine to prevent *M. tuberculosis* infections or provide a novel target for chemotherapy of tuberculosis.

5.2   Designing new drug-testing strategies for leprosy

Although there currently exists a number of chemotherapeutic agents to treat tuberculosis or leprosy, the long duration of treatment and the increase in the numbers of drug-resistant mutants necessitate the need to find novel drugs to treat mycobacterial disease. Infections caused by *M. avium* in patients with the Acquired Immunodeficiency Syndrome (AIDS) often go untreated as existing antibiotics are ineffective in adequately controlling these infections. Much basic research is needed to identify definitively the drug targets of existing drugs and the mechanisms of drug-resistance. The ability to manipulate genetically *M. smegmatis* may facilitate such analyses for *M. avium*, as well as for *M. tuberculosis* and *M. leprae*.

The potential usefulness of a genetic approach to identify more effective drugs can be illustrated with the search for novel anti-leprosy agents. To assay candidate anti-leprosy drugs or vaccines presently requires the incredibly slow and arduous procedure of inoculating *M. leprae* from tissue biopsies into the mouse footpad and waiting nine months until sufficient growth occurs to enable visual counting of bacilli. Drugs are administered during this long period of infection and the efficacy of these drugs is determined after nine months by their ability to diminish the number of *M. leprae* cells in mouse footpads. The genetic systems recently developed in mycobacteria, as described above, may provide alternative and more efficient means of drug testing. After identifying appropriate target enzymes for anti-mycobacterial drugs, it may be possible to isolate *M. leprae* genes encoding those enzymes by a variety of approaches. By then replacing the homologous genes in a rapidly growing cultivable *Mycobacterium* such as *M. smegmatis* with the *M. leprae* gene encoding the putative drug target (Figure 9.6), it

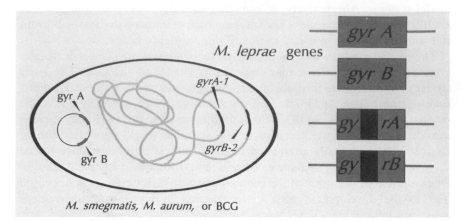

Figure 9.6 Recombinant mycobacteria for screening anti-leprosy drugs. Avirulent mycobacterial host strains can be used to express genes encoding *M. leprae* drug targets. In this diagram the target genes from *M. leprae* (*gyr*A and *gyr*B encoding DNA gyrase subunits) are cloned on a plasmid vector and expressed in a host strain whose endogenous analogous genes have been mutated (*gyr*A–1, *gyr*B–2).

should be possible to develop rapid (1–4 day) assays for screening novel drugs by their ability to inhibit growth of the recombinant test strain expressing that *M. leprae* enzyme.

6   DEVELOPING A RECOMBINANT BCG MULTIVACCINE VEHICLE

BCG presents unique advantages for developing a recombinant polyvalent vaccine vector expressing antigens for a wide variety of pathogens, particularly those for which cell-mediated immunity is important in protection (Bloom, 1986; Jacobs *et al.*, 1988). Such a vaccine could be achieved if genes encoding protective antigens from a variety of pathogens could be introduced and stably expressed in the BCG cell (Figure 9.7). Although the systems to introduce recombinant DNA into BCG have now been developed (Jacobs *et al.*, 1987; Snapper *et al.*, 1988), and the first foreign bacterial antigen — the 65 kD *M. leprae* heat-shock protein — has been expressed in BCG (S. B. Snapper, B. R. Bloom, and W. R. Jacobs Jr, manuscript in preparation), much work remains to be done to develop BCG into a useful multivaccine vehicle.

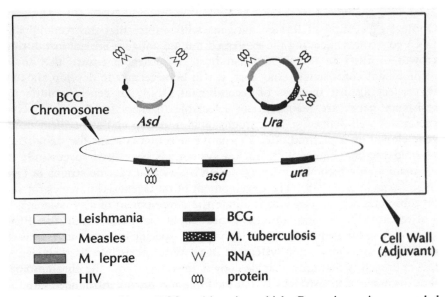

Figure 9.7   A recombinant BCG multivaccine vehicle. Protective antigens encoded from genes from a variety of pathogens are expressed from plasmid vectors containing wildtype markers (e.g. URA, ASD). These markers can complement chromosomal mutations (e.g. ura, asd) providing a means of plasmid selection.

6.1   Optimization of the expression of foreign antigens

In order to express any open reading frame of DNA derived from any foreign antigen in BCG it will be necessary to identify mycobacterial expression sequences. Mycobacteria contain high G–C DNA and, while related to Gram-positive organisms, they may have special DNA sequences useful for expressing foreign genes. A technique for isolating mycobacterial expression sequences, using the galactosidase reporter gene in *M. smegmatis* and BCG, has been recently developed (R. G. Barletta, D. Kim, B. R. Bloom, and W. R. Jacobs Jr, manuscript in preparation). The detailed characterization of these sequences should enable the construction of a series of expression vectors capable of expressing any open reading frame. The expression of a reporter gene in mycobacteria provides both a simple tool for identifying DNA fragments that mediate gene expression, and a tool for comparing the level of expression of different mycobacterial promoters to further identify and characterize factors involved in mycobacterial gene expression.

6.2   Stable maintenance of recombinant DNA in BCG

Optimal expression of foreign antigens will require that the recombinant DNA constructs encoding the expressed foreign antigens are not lost during growth of BCG in the mammalian host. In order to ensure that these recombinant constructs are not lost, it will be necessary to develop systems that select against the loss of recombinant DNA. In general, antibiotic resistance genes are inappropriate selectable markers for human vaccine vectors. A well-defined stable chromosomal mutation and a complementing gene cloned on a plasmid offer alternative selection systems for stablilizing plasmid vectors (Figure 9.7). Such a system has been used successfully to maintain stably recombinant plasmids in *Salmonella* vaccine strains *in vivo* (Nakayama *et al.*, 1988). The development of mutagenesis strategies for the mycobacteria are necessary to facilitate the development of such systems.

Alternatively, foreign antigen genes could be placed into the BCG chromosome by using mycobacteriophage site-specific integration systems or homologous recombination systems of BCG. While a possible disadvantage to this approach is that only a single copy is introduced into the chromosome, the advantage of this strategy is that the foreign genes are introduced into the host chromosome and thus may not require continuous selection.

Finally, the expression of each foreign antigen gene may be difficult because of protein degradation or cell toxicity as a result of antigen expression. It is clear that expression of some foreign genes in *E. coli* is limited by host enzymes capable of selectively degrading foreign proteins. It has been possible, however, to stabilize expression of foreign proteins in *E. coli* by developing mutants in proteolytic enzymes (Goldberg and St John, 1976). Similar approaches may be useful to increase the stability of expression of foreign antigens in BCG. Clearly, the development of an effective recombinant BCG multivaccine, containing protective antigens from a variety of infectious agents and capable of eliciting a protective immune response to each pathogen, will require further investigation of gene expression in BCG in order to determine the optimal conditions necessary to express each foreign antigen.

REFERENCES

Baess, I. (1970). In *Host–Virus Relationships in* Mycobacterium, Nocardia, *and* Actinomyces (eds S. E. Juhasz and G. Plummer), pp. 210–14. C. C. Thomas, Springfield.
Baess, I. (1971). *Acta Path. Microbiol. Scand. Section B*. **79**, 428–34.
Bibb, M. J., Ward, J. M. and Hopwood, D. A. (1978). *Nature* **274**, 398–400.
Bloom, B. R. (1986). *J. Immunol.* **137**, i–x.
Clark-Curtiss, J. E., Jacobs, W. R., Docherty, M. A., Ritchie, L. R. and Curtiss, R., III. (1985). *J. Bacteriol.* **161**, 1093–102.

Collins, J. and Hohn, B. (1978). *Proc. Natl. Acad. Sci. USA* **75**, 4242–6.

Crawford, J. T. and Bates, J. H. (1979). *Infect. Immun.* **24**, 979–81.

Crawford, J. T. and Bates, J. H. (1984). *Gene* **27**, 331–3.

Crawford, J. T., Cave, M. D. and Bates, J. H. (1981a). *J. Gen. Microbiol.* **127**, 333–8.

Crawford, J. T., Cave, M. D. and Bates, J. H. (1981b). *Rev. Infect. Dis.* **3**, 949–52.

Doke, S. (1960). *Kumamoto Med. J.* **34**, 1360–73.

Fields, P. I., Swanson, R. V., Haidaus, C. G. and Heffron, F. (1986). *Proc. Natl. Acad. Sci. USA* **83**, 5189–93.

Fry, K. L., Scott Meissner, P. and Falkinham, J. O., III. (1986). *Am. Rev. Respir. Dis.* **134**, 39–43.

Gicquel-Sanzey, B., Moniz-Pereira, J., Gheorghiu, M. and Rauzier, J. (1989). *Acta Leprologica* **7** (Suppl 1), 208–11.

Goguen, J. D., Yother, J. and Straley, S. C. (1984). *J. Bacteriol.* **160**, 842–8.

Goldberg, A. L. and St John, A. C. (1976). *Annu. Rev. Biochem.* **45**, 747–803.

Greenberg, J. and Woodley, C. L. (1984). In *The Mycobacteria: A Sourcebook* (eds G. P. Kubica and L. G. Wayne) pp. 629–39. Marcel Dekker, New York.

Hopwood, D. A. and Wright, H. M. (1978). *Molec. Gen. Genet.* **162**, 307–17.

Hopwood, D. A., Bibb, M. J., Chater, K. F., Kieser, T., Bruton, C. J., Kieser, H. M., Lydiate, D. J. and Smith, C. P. (1985). *Genetic Manipulation of* Streptomyces: *A Laboratory Manual.* John Innes Foundation, Norwich.

Hull, S. I., Wallace, R. J., Bolbey, D. G., Price, K. E., Goodhines, R. A., Swenson, J. M. and Silcox, V. A. (1984). *Am. Rev. Resp. Dis.* **129**, 614–18.

Isberg, R. R. and Falkow, S. (1985a). *Curr. Top. Microbiol.* **118**, 1–11.

Isberg, R. R. and Falkow, S. (1985b). *Nature* **317**, 262–4.

Jacobs, W. R., Docherty, M. A., Curtiss, R., III. and Clark-Curtiss, J. E. (1986). *Proc. Natl. Acad. Sci. USA* **83**, 1926–30.

Jacobs, W. R., Jr, Tuckman, M. and Bloom, B. R. (1987). *Nature* **327**, 532–5.

Jacobs, W. R., Jr, Snapper, S. B. and Bloom, B. R. (1988). In *Molecular Biology and Infectious Diseases* (ed. M. Schwartz) pp. 207–12. Elsevier, New York.

Jacobs, W. R., Jr, Snapper, S. B., Tuckman, M. and Bloom, B. R. (1989). *Rev. Infect. Dis.* **11**, S404–10.

Khandekar, P. S., Munshi, A., Sinha, S., Sharma, G., Kapoor, A., Gaur, A. and Talwar, G. P. (1986). *Int. J. Leprosy* **54**, 416–22.

Kieser, T. and Melton, R. E. (1988). *Gene* **65**, 83–103.

Kieser, T., Moss, M. T., Dale, J. W. and Hopwood, D. A. (1986). *J. Bacteriol.* **168**, 72–80.

Kitahara, K. and Sellers, M. I. (1975). *Ann. Sclavo.* **17**, 605–6.

Labidi, A., Dauguet, C., Goh, K. S. and Gavid, H. L. (1984). *Curr. Microbiol.* **11**, 235–40.

Labidi, A., David, H. L. and Roulland-Dussoix, D. (1985). *FEMS Microbiol. Lett.* **30**, 221–5.

Lugosi, L., Jacobs, W. R., Jr and Bloom, B. R. (1989). *Tubercle* **70**, 159–70.

Maurelli, A. T. and Curtiss, R., III. (1984). *Infect. Immun.* **45**, 642–8.

Maurelli, A. T., Baudry, B., d'Hauteville, H., Hale, T. L. and Sansonetti, P. J. (1985). *Infect. Immun.* **49**, 164–71.

Menezes, J. and Pavilanis, V. (1969). *Experientia* **25**, 1112–13.

Mizuguchi, Y. and Tokunaga, T. (1968). *Med. Biol. (Tokyo)* **77**, 57–60.

Mizuguchi, Y., Fukunaga, M. and Taniguchi, H. (1981). *J. Bacteriol.* **146**, 656–9.

Mizuguchi, Y., Udou, T. and Yamada, T. (1983). *Microbiol. Immunol.* **27**, 425–31.

Nakamura, R. M. (1970). In *Host–Virus Relationships in* Mycobacterium, Nocardia, *and* Actinomyces. (eds S. E. Juhasz and G. Plummer) pp. 166–78. C. C. Thomas, Springfield.

Nakayama, K., Kelly, S. M. and Curtiss, R., III. (1988). *Bio/Tech* **6**, 693–7.

Okanishi, M., Suzuki, K. and Umezawa, H. (1974). *J. Gen. Microbiol.* **80**, 389–400.

Rauzier, J., Moniz Pereira, J. and Gicquel Sanzey, B. (1988). *Gene* 71, 315–21.

Rieber, M. and Imaeda, T. (1970). In *Host–Virus Relationships in* Mycobacterium, Nocardia, *and* Actinomyces (eds S. E. Juhasz and G. Plummer, pp. 144–51. C. C. Thomas, Springfield.

Rosenberg, S. M., Stahl, M. M., Kobayashi, I. and Stahl, F. W. (1985). *Gene* 38, 165–75.

Scott Meissner, P. and Falkinham, J. O., III. (1984). *J. Bacteriol.* 157, 669–72.

Scott Meissner, P. and Falkinham, J. O., III. (1986). *J. Infect. Dis.* 153, 325–31.

Sellers, M. I. and Tokunaga, T. (1966). *J. Exp. Med.* 123, 327–40.

Sellers, M. I., Nakamura, R. and Tokunaga, T. (1970). *J. Gen. Virol.* 7, 233–47.

Snapper, S. B., Lugosi, L., Jekkel, A., Melton, R. E., Kieser, T., Bloom, B. R. and Jacobs, W. R. (1988). *Proc. Natl. Acad. Sci. USA* 85, 6987–91.

Snapper, S. B., Melton, R. E., Kieser, T. and Jacobs, W. R., Jr (1990). Manuscript submitted.

Taylor, R. K., Manoil, C. and Mekalanos, J. J. (1989). *J. Bacteriol.* 171, 1870–8.

Thole, J. R., Dauwese, H. G., Das, P. K., Groothuis, D. G., Schouls, L. L. M. and van Embden, J. D. A. (1985). *Infect. Immun.* 50, 800–6.

Timme, T. L. and Brennan, P. J. (1984). *J. Gen. Microbiol.* 130, 2059–66.

Tokunaga, T. and Nakamura, R. M. (1967). *J. Virol.* 1, 448–9.

Tokunaga, T. and Nakamura, R. M. (1968). *J. Virol.* 2, 110–17.

Tokunaga, T. and Sellers, M. I. (1964). *J. Exp. Med.* 119, 139–49.

Tokunaga, T. and Sellers, M. I. (1970a). In *Host–Virus Relationships in* Mycobacterium, Nocardia, *and* Actinomyces (eds S. E. Juhasz and G. Plummer) pp. 152–65. C. C. Thomas, Springfield.

Tokunaga, T. and Sellers, M. I. (1970b). In *Host–Virus Relationships in* Mycobacterium, Nocardia, *and* Actinomyces (eds S. E. Juhasz and G. Plummer) pp. 227–43. C. C. Thomas, Springfield.

Udou, T., Ogawa, M. and Mizuguchi, Y. (1982). *J. Bacteriol.* 151, 1035–9.

Udou, T., Ogawa, M. and Mizuguchi, Y. (1983). *Can. J. Microbiol.* 29, 60–8.

Weiss, A. A., Herskett, E. L., Myers, G. A. and Falkow, S. (1983). *Infect. Immun.* 42, 33–41.

Young, R. A., Bloom, B. R., Grosskinsky, C. M., Ivanyi, J., Thomas, D. and Davis, R. W. (1985a). *Proc. Natl. Acad. Sci. USA* 82, 2583–7.

Young, R. A., Mehra, V., Sweetser, D., Buchanan, T., Clark-Curtiss, J., Davis, R. W. and Bloom, B. R. (1985b). *Nature* 316, 450–2.

Zainuddin, Z. F., Kunze, Z. M. and Dale, J. W. (1989a). *Molecular Microbiology* 3, 29–34.

Zainuddin, Z. F., Kunze, Z. M. and Dale, J. W. (1989b). *Acta Leprologica* 7 (Suppl 1), S212–16.

10 Mycobacterial molecular biology and genetics: challenges for the future

Richard A. Young

*Whitehead Institute for Biomedical Research, Nine Cambridge
Center, Cambridge, MA 02142, USA
and
Department of Biology, Massachusetts Institute of Technology,
Cambridge, MA 02139, USA*

1 THE PROBLEM

In his book *Plagues and Peoples* historian William McNeill argues that infectious diseases have had an enormous impact on human history (McNeill, 1976). Tuberculosis and leprosy are among those diseases that have had, and that continue to have, a significant impact on mankind. The World Health Organization (WHO) estimates that one of every three people in the world are infected with the bacillus that causes tuberculosis, that approximately 10 million new cases of tuberculosis occur annually, and that 2–3 million people die from tuberculosis each year. Leprosy afflicts some 10 million individuals and, although it is rarely life threatening, the disease irreparably damages nerves and frequently alters the lives of the afflicted (Bloom, 1989).

Over the last decade, an international group of molecular biologists and immunologists have been investigating the molecular biology and immunology of the mycobacteria to understand better these pathogens and to obtain information that might be useful in the fight against these diseases. This work has led to important insights into the immune response to infection by mycobacteria and other pathogens, to knowledge of some of the basic molecular features of the mycobacteria, and to the development of techniques

MOLECULAR BIOLOGY OF THE MYCOBACTERIA
ISBN 0–12–483378–0

and tools for the molecular genetic manipulation of mycobacterial DNA. The future of research into mycobacterial molecular biology and genetics should reveal much about these pathogens, but there are some challenges particular to mycobacteria and the diseases that they cause that need to be met.

2 THE PAST AND THE PRESENT

Genes that encode a dozen or so mycobacterial protein antigens have been isolated and characterized by probing recombinant DNA expression libraries of mycobacterial DNA with antibody or T-cell probes. Many of these antigens turn out to be heat-shock proteins (Baird *et al.*, 1988; Shinnick *et al.*, 1988, 1989; Young *et al.*, 1988; Thangaraj *et al.*, 1989; Young, 1990; see D. Young *et al.*, this volume). This discovery led to the observation that members of stress protein families are prominent antigens in a diverse range of bacterial and parasitic infections, and workers are now tryng to assess whether they play a role in immune protection. Mycobacterial antigens that are not stress proteins have also been identified. The antibody and T-cell response mounted against most of these mycobacterial proteins is being examined further to determine whether any of these proteins show promise as candidates for subunit vaccines.

There have been equally important observations on the nature of the mycobacterial genome, revealing features unique to mycobacterial DNA and providing insights into phylogenetic relationships between mycobacteria and other organisms. The observation that *Mycobacterium bovis* BCG contains only one or two ribosomal RNA cistrons (Suzuki *et al.*, 1987) may help account for the ability of some mycobacteria to maintain slow growth; this may be important if the maintenance of slow growth is crucial for pathogenesis. Fast-growing bacteria often have at least half a dozen rRNA cistrons, and devote half of the RNA synthesis apparatus to the synthesis of rRNA. DNA sequence comparisons have substantially improved our understanding of the phylogenetic relationships between mycobacteria, in particular the pathogenic mycobacteria. In addition, the identification of species-specific DNA sequences and the application of efficient techniques to probe for them may help efforts to improve the tools available for mycobacterial detection and for epidemiological studies (see McFadden, this volume).

Methods and tools that permit transformation and molecular genetic manipulation of mycobacteria have recently been developed (Jacobs *et al.*, 1987; Snapper *et al.*, 1988; Husson *et al.*, 1990). Shuttle vectors are available that permit the manipulation of recombinant DNA in *E. coli* and the introduction of the recombinant DNA into mycobacteria. These tools should provide new approaches to investigate mycobacterial biology and pathogenesis and a means to develop mycobacteria into live recombinant vaccine vehicles.

3 FUTURE PROSPECTS

What are the challenges for mycobacterial research in the future? There are immediate challenges and long-term challenges. Among the immediate problems are determining whether any of the mycobacterial proteins can provide immunological protection against mycobacterial infection in animal models; improving the tools available for mycobacterial diagnosis; developing molecular and genetic approaches to identify potentially useful drugs; and improving the tools and methods for mycobacterial molecular genetics. Among the long-term problems, we must determine whether BCG can be an effective live recombinant vaccine vehicle; ascertain whether one can improve upon BCG for tuberculosis immunoprophylaxis; and develop an effective and inexpensive leprosy vaccine.

4 ANTIGENS FOR VACCINE CANDIDATES

A substantial number of mycobacterial genes that encode antigens have been isolated and are in various stages of characterization. To investigate the immune response to specific antigens and to determine whether any of the proteins have immunoprophylactic properties in animal models, investigators need large amounts of the antigens of interest, especially if they are to compare the properties of one protein antigen with another. To help facilitate these studies, the Tropical Disease Research programme of WHO is supporting the production and purification of many of the mycobacterial protein antigens whose genes have been well characterized, and some of these recombinant proteins are already available to qualified investigators. In addition, vaccinia recombinants that express a variety of mycobacterial proteins have been constructed by Applied bioTechnology, Inc. (Cambridge, Massachusetts), and these recombinant viruses are also available to qualified investigators.

There is no reason to believe that any one protein antigen will be as effective as a combination of antigens in immunoprophylaxis for large human populations. Investigators should expect that more than one mycobacterial protein will contribute to the efficacy of a subunit vaccine. Thus, collaborative efforts to identify a collection of antigens that may be useful for immune protection are most likely to succeed in producing effective vaccine candidates. This may be a challenge for investigators who use animal models to study immune protection, as many different candidates need to be tested, perhaps both alone and in combination with one another.

5 DIAGNOSIS OF MYCOBACTERIA

There continues to be a need to improve the tools available for mycobacterial diagnosis. For tuberculosis, an inexpensive test that can rapidly identify cases of clinical disease is needed. Such a test would have a substantial impact, as a single infectious individual is a source of infection for large numbers of people. For leprosy, where the bacillus can be cured but nerve damage is irreversible, early detection is especially important. Perhaps a test that can differentiate between viable and inviable *M. leprae* will help, although this problem seems especially challenging. It is conceivable that the Polymerase Chain Reaction (PCR) technique might be exploited to this end.

6 IDENTIFICATION OF NEW DRUGS

New and improved drugs are always welcome. Drug resistance is a major problem in many tuberculosis and leprosy endemic areas, and multidrug therapy is recommended. Molecular and genetic approaches could conceivably be developed to identify potentially useful drug candidates. For example, mycobacterial genes that encode proteins that are potential targets for drug action can be cloned and used to produce enzymes in a form useful for drug screening. Among the enzymes that might be useful are RNA polymerase, DNA gyrase and dihydrofolate reductase.

7 MYCOBACTERIAL MOLECULAR GENETICS

A physical map of the mycobacterial genome would be very useful for a variety of reasons. It would be a resource of genetic information that would permit mycobacterial molecular geneticists to determine most efficiently the organization of genes on the bacterial chromosome. It would facilitate comparisons of genome organization among mycobacterial species, and may help identify probes useful for specific diagnosis and epidemiology. Moreover, it would make mycobacteria more attractive to molecular geneticists, especially for young investigators who are looking for a good molecular genetic system for study in the field of infectious diseases. Indeed it seems appropriate seriously to consider sequencing an entire mycobacterial genome.

Mycobacteria can be stably transformed with recombinant DNA. There are now a limited number of vectors that replicate autonomously in *M.*

smegmatis and in *M. bovis* BCG, or that permit homologous integration and gene replacement in *M. smegmatis* genomic DNA (Jacobs *et al.*, 1987; Snapper *et al.*, 1988; Husson *et al.*, 1990). Mycobacterial mutants have been constructed using homologous integration with recombinant mycobacterial DNA. These tools and methods make it possible to begin manipulating the mycobacterial genome in very precise ways. For example, specific mutations can now be introduced into any gene of interest and the phenotype of the resulting mutant can be investigated.

There are some serious experimental limitations to mycobacterial molecular genetics that have yet to be overcome. It has not yet been possible to obtain homologous integration in the BCG genome. This is important both because homologous integration is essential for genomic manipulation and because genomic integration may be essential for stable expression of foreign genes in BCG. This problem is likely to be solved through the development of more efficient transformation methods. Another serious limitation is the very limited number of genetic markers available in the current vectors. In particular, there is a need for nutritional markers; drug resistance markers are not useful for BCG recombinants destined for human use. Yet another problem is with the limited amount of information that is currently available on the nature of mycobacterial gene expression signals, both transcriptional and translational.

8   LONG-TERM CHALLENGES: BCG MAY BE A USEFUL VACCINE VEHICLE

BCG has a number of features that make it the most attractive live recombinant vaccine vehicle. As a vaccine against tuberculosis, it already has a long record of safe use in man; nearly 2 billion doses have been given to humans. BCG is an extremely effective adjuvant and engenders long-lived immune responses with only a single dose. Several additional features of BCG suggest that it would make an especially good live recombinant vaccine vehicle for the developing world. Here there are problems with the cost of vaccines, the stability of vaccines in the absence of refrigeration, and the delivery of vaccines where there are limited numbers of health professionals and access to the population is difficult. BCG is easy and inexpensive to manufacture; the current cost for the vaccine is 5.5 cents/dose. BCG is heat stable, and thus can be transported in areas of the world that lack facilities for refrigeration. There is already a world-wide distribution network set up with experience in BCG vaccination. BCG is easy to inoculate, and may be made easier by oral vaccination. It is one of the few vaccines that can be given at birth.

Will recombinant BCG work? If it does not work initially, are there modifications that might make it work? One of the attractive features of bacterial molecular genetics is that it offers almost unlimited possibilities to correct deficiencies with the design of a live recombinant vehicle vaccine. If the level of transcription of a gene is too low or too high, the promoter can be altered. If the rate of degradation of a protein antigen is too great or too little, the level of some proteases can be adjusted. Even the survival lifetime of the bacillus after inoculation might be fine-tuned. Some of these parameters will probably be different for different vaccines, and perhaps each new vaccine will provide new challenges.

9 CAN BCG BE IMPROVED?

The ability of BCG to protect against tuberculosis has varied considerably in a number of human vaccine trials (Luelmo, 1982). The failure of BCG to protect against tuberculosis in specific settings has been attributed to a number of factors, including the source of the vaccine strain, the vaccine dose, and the level of environmental mycobacteria. Given the complexity of vaccines, and our lack of understanding of precisely how they work, it is unclear why BCG has failed to provide protection in specific trials. Nonetheless, it is possible that the immunoprophylactic effect of BCG could be enhanced by specifically increasing the expression of genes whose products are important for immune protection. Once the antigens that are important for protection are identified, it is possible that BCG could be manipulated to increase the production of specific antigens.

One major challenge to improving vaccines will be the ability of animal models to provide relevant information about the relative efficacy of vaccine candidates. It seems likely that animal models will become a bottleneck as sizeable numbers of vaccine candidates are developed.

10 PERSPECTIVES

Our understanding of human diseases will increase enormously in the 1990s, in large part due to relatively recent advances in molecular biology and immunology. Improved understanding of mycobacterial molecular biology, genetics and immunology is likely to play an important role in helping the fight against mycobacterial diseases, both by providing basic information and by providing diagnostic and immunoprophylactic tools. With well-developed molecular and genetic tools, it is possible that advances in mycobacterial research will have a leading role in producing a better understanding of mycobacterial and other infectious diseases.

REFERENCES

Baird, P. N., Hall, L. M. C. and Coates, A. R. M. (1988). *Nucleic Acid Res.* **16**, 9047.

Bloom, B. R. (1989). *Nature* **342**, 115–20.

Husson, R. N., James, B. E. and Young, R. A. (1990). *J. Bacteriol.* **172**, 519–24.

Jacobs, W. R., Jr, Tuchman, R. and Bloom, B. R. (1987). *Nature* **327**, 532–5.

Luelmo, F. (1982). *Am. Rev. Respir. Dis.* **125**, 70–2.

McNeill, W. (1976). *Plagues and Peoples*, Anchor Press/Doubleday, Garden City, NY.

Shinnick, T. M., Vodkin, M. H. and Williams, J. C. (1988). *Infect. Immun.* **56**, 446–51.

Shinnick, T. M., Plikaytis, B. B., Hyche, A. D., Van Landingham, R. M. and Walker, L. L. (1989). *Nucleic Acids Res.* **17**, 1254.

Snapper, S. B., Lugosi, L., Jekkel, A., Melton, R. E., Kieser, T., Bloom, B. R. and Jacobs, W. R. Jr (1988). *Proc. Natl. Acad. Sci. USA* **85**, 6987–91

Suzuki, Y., Yoshinaga, K., Ono, Y., Nagata, A. and Yamada, T. (1987). *J. Bacteriol.* **169**, 839–43.

Thangaraj, H. S., Lamb, F. I., Davis, E. O. and Colston, M. J. (1989). *Nucleic Acids Res.* **17**, 8378.

Young, D. B., Lathigra, R., Hendrix, R., Sweetser, D. and Young, R. A. (1988). *Proc. Natl. Acad. Sci. USA* **85**, 4267–70.

Young, R. A. (1990). *Annu. Rev. Immunol.* **8**, 401–20.

REFERENCES

Index

vaccines in the world — and one of the safest — the development of BCG into a multivaccine vehicle is an important goal (Bloom, 1986; Jacobs *et al.*, 1988). A prerequisite, however, to the development of a gene transfer system is an efficient method for introducing DNA into mycobacteria.

2 MYCOBACTERIAL TRANSFORMATION

While there have been some successful reports of transformation in mycobacteria using chromosomally encoded markers, a large number of similar experiments have failed (reviewed by Greenburg and Woodley, 1984). Because of the difficulty in obtaining consistent results with chromosomally encoded markers and the unavailability of useful selectable markers on plasmids for transforming mycobacteria, another strategy was chosen to evaluate optimum conditions for DNA introduction into mycobacteria, the transfection of DNA from lytic mycobacteriophages. The advantages of this assay are that the results obtained are quantitative and plaques are readily visible within 24 hours. Tokunaga and Sellers (1964) reported first the introduction of mycobacteriophage DNA into whole *M. smegmatis* cells. These studies were later confirmed by several investigators (Sellers and Tokunaga, 1966; Mizuguchi and Tokunaga, 1968; Tokunaga and Nakamura, 1968; Nakamura, 1970; Rieber and Imaeda, 1970; Sellers *et al.*, 1970; Tokunaga and Sellers, 1970a) with DNA from other phages and by other investigators (Tokunaga and Nakamura, 1967; Menezes and Pavilanis, 1969; Kitahara and Sellers, 1975) who extended these observations to other species of mycobacteria. In these studies, various cell preparations were examined, including cells grown in broths of an acid pH (Nakamura, 1970), and broths containing divalent cations (Tokunaga and Sellers, 1964; Tokunaga and Nakamura, 1967; Tokunaga and Nakamura, 1968; Rieber and Imaeda, 1970; Sellers *et al.*, 1970), glycine and lysozyme (Mizuguchi and Tokunaga, 1968), or amino acids (Mizuguchi and Tokunaga, 1968; Tokunaga and Sellers, 1970a). However, the efficiency of transfection was discouraging, less than 100 pfu/µg. Jacobs *et al.* (1987) extended the transfection protocols of Mizuguchi and Tokunaga (1968) by adapting the technology of protoplast preparation from *Streptomyces* (Okanishi *et al.*, 1974; Hopwood and Wright, 1978) for mycobacterial cells with the addition of polyethylene glycol to promote DNA entry (Bibb *et al.*, 1978) (See Figure 9.1). Transfection efficiencies of greater than 10^4 plaque-forming units (pfu) were achieved. This work was recently confirmed elsewhere (Gicquel-Sanzey *et al.*, 1989). While the regeneration of mycobacterial protoplasts to bacillary forms has been described (Udou *et al.*, 1982, 1983), the efficiency of this process is unclear and some attempts to introduce plasmids containing selectable

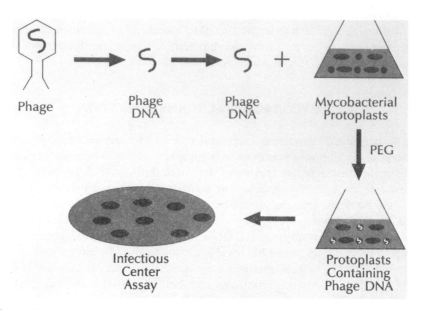

Figure 9.1 Mycobacterial transfection strategy. DNA, isolated from mycobacteriophage particles, is combined with mycobacterial protoplasts. Polyethylene glycol (PEG) is added to this suspension to promote DNA entry. DNA entry is monitored by plaque formation in an infectious centre assay.

markers into mycobacterial protoplasts have failed (unpublished results, S. B. Snapper and W. R. Jacobs Jr). Consequently, the first approach to develop a mycobacterial gene transfer system was to construct recombinant DNA vectors from mycobacteriophages.

3 PHAGE-BASED GENE TRANSFER SYSTEMS

Phages were excellent candidates for development as recombinant DNA vectors, since they efficiently infect and replicate in a wide range of fast-growing and slow-growing mycobacterial species. Moreover, successful phage-based genetic systems would not require the regeneration of proto-plasts since whole, untreated, sensitive cells are included in the transfection assay. The construction of a recombinant mycobacteriophage vector, as in the construction of any phage vector, requires the introduction of a foreign fragments(s) of DNA within the phage genome while still maintaining a viable phage that will replicate the cloned sequences. It is therefore impera-tive to have regions of the phage genome which are nonessential for